T0168626

Science or Pseudoscience

Magnetic Healing, Psychic Phenomena, and Other Heterodoxies

Science or Pseudoscience

Magnetic Healing, Psychic Phenomena, and Other Heterodoxies

Henry H. Bauer

University of Illinois Press
Urbana and Chicago

Manufactured in the United States of America
∞ This book is printed on acid-free paper.

Library of Congress Cataloging-in-Publication Data
Bauer, Henry H.
Science or pseudoscience : magnetic healing, psychic phenomena, and other heterodoxies / Henry H. Bauer.
p. cm.
Includes bibliographical references and index.
ISBN 0-252-02601-2 (cloth : alk. paper)
1. Science—Miscellanea. 2. Belief and doubt.
3. Parapsychology and science. 4. Skepticism.
I. Title.
Q157.B24 2001
133—dc21 00-008332

C 5 4 3 2 1

Every folly has its reason, which it is never amiss to understand.
—Jacques Barzun, *Science: The Glorious Entertainment*

In all controversies, it is better to wait the decisions of time, which are slow and sure, than to take those of synods, which are often hasty and injudicious.
—Joseph Priestley, *The Rudiments of English Grammar*

People who never chase after wild herrings or red geese rarely stand the world on its ear.
—attr. Doc Morgan Nelson, in Christine McGuire's *Until Proven Guilty*

Contents

Preface

Many fascinating topics are ignored by science: psychic phenomena, UFOs, living man-apes (Bigfoot, yeti), prehistoric alien visitors to Earth—to name just a few of them.

Some pundits and science groupies call these things "pseudoscience" and poke fun at people who take them seriously. But if you ask how you can tell when something is pseudoscience, you're unlikely to get a straight answer. Probably you'll hear something about not using the scientific method or perhaps about the need for theories to be falsifiable (as though there were something wrong with believing what cannot be disproved). The supposed experts on this, the philosophers of science, in their honest moments will admit that no satisfactory definition of pseudoscience exists. "Pseudoscience," it turns out, is less an intellectual category than an epithet.

Just as those who use the n-word reveal racism, so do those who sneer at "pseudoscience" reveal scientism: the belief that only science is authoritative when it comes to knowledge.

One trouble with that attitude is that a great deal *within science* is *incorrect* "knowledge." The history of science is a continuing chronicle of old ideas discarded, theories superseded, "facts" modified.

Another trouble is that some things *within science itself* get called "pseudoscience" even though they are proposed and taken seriously by accomplished scientists: things like cold fusion and polywater.

But perhaps the chief trouble is that some of these startling, outrageous "pseudoscientific" claims turn out to have genuine substance to them, as happened with meteorites a couple of centuries ago or acupuncture a few decades ago or ball lightning even more recently.

Wescott (1973, 1980) has called "anomalistics" the study of things that science denigrates or ignores, that are often labeled "pseudoscience." In this book I make a detailed comparison of the actual practices in anomalistics and in scientific work. Thereby it becomes evident that there is no easy or sharp division to be made between science and anomalistics, particularly since "science" must be understood to include the social as well as the natural sciences.

The fact of the matter is that extending humankind's knowledge is desperately difficult and fraught with pitfalls. Being within science or being an eminently competent and accomplished scientist is insufficient safeguard against falling into such a pit. Though the progress of science may seem smooth and steady to outsiders, its magnificent accomplishments over the last few centuries have actually been accompanied by all sorts of false trails and errors that get corrected only by continuing effort.

Since science as well as "pseudoscience" suffers from mistakes, it makes little sense to attempt to distinguish them on grounds of what is true or not. As to casting aspersions at unorthodox claims, one might usefully distinguish between serious anomalistics—unorthodox knowledge pursued by genuine seekers of new truths—and intellectual trash, fraudulent so-called "knowledge" dogmatically peddled by people who seem primarily to wish to gain personally thereby.

Harm can come to individuals and even to society, as the inquisitors of "pseudoscience" often allege, from buying the nostrums of the indiscriminate, unscrupulous, and often dishonest "knowledge"-peddlers. *Serious* anomalistics, on the other hand, can be intellectually and emotionally rewarding for the individuals engaged in it, and it can also yield results beneficial to society as a whole.

Exploring mysteries in the realm of anomalistics, as in that of science, brings home the vastness of human ignorance. Thus about the connections between electromagnetism and living processes we know only enough to be sure that there's a lot remaining to be understood. What we've learned about "chaos" and "complexity" is less solid knowledge than a recognition that we still have much to learn about how complex systems work.

The precursors of modern science, the practitioners of natural history, often displayed an entirely appropriate humility, a worshipful attitude toward the wonders of nature that they sought to describe and understand. During the nineteenth century, the staggering achievements of the natural sciences generated a widespread hubris—by no means confined within the scientific community—that physics, chemistry, and the other sciences are capable of finding good answers to every question human beings care to ask, even including the meaning and purpose of human existence. That *scientistic* belief, that scientistic illusion, seems ready to be discarded. The pursuit of anomalistics is in one sense a return to the approach taken in natural history, an exploration of the endless realm of the unknown. Like science, anomalistics can be a "glorious entertainment" (Barzun 1964); and a comparison of anomalistics with science can also illuminate the nature of science and of knowledge seeking in general.

Acknowledgments

Many people indeed have helped me over the years in ways relevant to this book.

Jack (I. J.) Good has been an unparalleled resource on all matters intellectual as well as statistics in particular. E. C. Krupp shared his considerable insight into archaeoastronomy. Leroy Ellenberger freely helped with his unrivaled knowledge of Velikovskiana and other recondite matters. Joel Carlinsky stimulated me to learn and think more about Wilhelm Reich and orgonomy and drew to my attention the Origin-of-Life Prize. Cliff Bryant guided me to the sociological literature on deviance. Richard D. Smith has been a valued discussant of matters as varied as cryptozoology and bluegrass music. Dónal O'Mathúna led me to fascinating material on alternative medicine.

Marcello Truzzi and Ron Westrum were early mentors about anomalistics. Marcello introduced me to Peter Sturrock when the Society for Scientific Exploration was being formed, which brought me into contact with many other sterling people. I've learned a great deal from and through them.

I'm truly grateful also to the many colleagues at Virginia Polytechnic Institute and State University who have been my mentors in science studies (in its substantive, non-postmodernist, aspects, one perhaps needs to say nowadays). I'm grateful to the many students over the years who taught me many things and caused me to learn even more. And I'm indebted to the several administrators who allowed me to go adventuring intellectually.

Science or Pseudoscience

Magnetic Healing, Psychic Phenomena, and Other Heterodoxies

1 Defining Categories

"Warm water freezes faster than cold," he alleged. He'd heard it, more or less accepted it, but never bothered to test it. Testing was easy enough. The claim proved to be wrong—as I had asserted it *must* be, for on the way to freezing, the warm water would first have to become cold, and that takes (extra) time.

"Eggs can be stood on end at the time of the equinox," one of my students told me. Not only had he heard that, he'd actually tried it and found it to be true. To prove it he brought photos taken at equinox time. "Could have been glue on the egg," I responded. To combat my skepticism, at the next equinox he brought a couple of eggs to class and demonstrated. For the next week or two, my family and I were much engaged in trying to stand eggs on end. We found that, given some persistence, nearly any egg can be stood on its larger end, equinox or not. That student had tested the claim, but he had not tested all of the claim; he had not done control experiments at other times than the equinox.

"Police radar can't pick up your car if you put aluminum foil in the hubcaps," other students informed me; and "You can beat a breathalyzer test by putting a copper penny under your tongue." Those claims were not so easily testable, so I called the state police laboratory for help. Technicians there were familiar with both claims and assured me that they are not valid.

The faster you go, the slower time passes and the heavier you get. Nothing in our everyday experience supports that claim. Yet, our best minds assure us, it is so and we should all believe it, though the effect is appreciable only at speeds that submicroscopic particles but not human beings can experience.

Nuclear fusion can be made to generate energy in a test tube at ordinary temperatures, by using palladium electrodes to pass electricity through "heavy" water. Some very competent people believe that. Some equally competent people dispute it.

There's a similar division of expert opinion over whether we're experiencing global warming because of the fossil fuels we've been burning; whether we've finally discovered the first planet outside our own solar system; whether we shall soon be able to make measurements of gravity waves; and a host of other scientific subjects.

Some people can predict the future. Some people can make contact with deceased loved ones. Some people have been abducted by aliens in UFOs and physically examined before being returned home. Those and many similar claims are made by intelligent people of considerable accomplishment. The same claims are dismissed as superstition or pseudoscience by other intelligent people of considerable accomplishment.

Which authorities or experts are we to believe? Is it possible to find out whether such a claim is true or not? Is there at least some way to form a reasonable opinion?

If scientific knowledge were the only knowledge important to human beings—which it is not—then one could simply turn to science for guidance on what to believe.

If scientific knowledge were always correct—which it is not—then one would know what to believe at least on those matters with which science concerns itself.

If science were easily defined—which it is not—then it would also be easy to know when claimed knowledge is *pseudo*science rather than the real thing.

But as things stand, there is available no quick or easy guidance about what to believe, not only on the many matters over which apparently competent people differ but also over some where the experts seem to be in agreement. At times we do well to believe what we're told; at other times we had better not. Sometimes there's no better guide than the experience of what you've seen for yourself; at other times your eyes deceive you. We should be open to new ideas—but on the other hand we should always be skeptical and critical before accepting a new idea, for old beliefs are often well tested by experience whereas new ones may just be untested hunches. It's good to see the whole picture, to be holistic, to be interdisciplinary—but on the other hand, in many fields progress requires concentration on ultraspecialized techniques, theories, and facts.

One of the most pervasive generalities encountered in arguments over beliefs is that one ought to turn to science, or to the scientific method, to get a reliable answer. But many bits of erstwhile science turn out to be wrong; and some things that are true don't need to be and can't be scientifically proved, for example, that I love my daughters, and indeed science has nothing useful to say about such things.

The success of science over the last few centuries has been so impressive that we've come to equate science globally with truth. So when there are arguments about extrasensory perception or astrology or the like, it seems natural to ask what science has to say about it. It also seems to offer a desirable shortcut: instead of arguing directly over the particular evidence, we just invoke the authority of science. But where unorthodox claims persist for a long time, as, say, with psychic phenomena, this appeal to science fails to settle the matter. Instead of a shortcut, it turns out to be a distraction, a red herring. We argue over whether or not extrasensory perception or the like is science instead of whether or not it happens.

Knowledge fights are really about beliefs. It seems natural to presume that the things we believe are in fact true; indeed we imagine that we believe them *because* they're true. But of course that's not always the case—if it were, disagreements might be rare, whereas in fact they're commonplace. Arguing over whether or not something is "science," whether or not it is "pseudoscience," often is an attempt to enlist science's supposedly objective knowledge in support of personal beliefs.

Criteria for what might be said to be true beliefs, or for distinguishing science from pseudoscience, have been disputed for at least as long as we have written records. Still philosophy has not arrived at an answer on which all agree. For most of us and as far as everyday life is concerned, those are truly academic arguments. If something happens every time we try it, if there has been no exception in a large number of repetitions, that makes it true enough for our everyday purposes. If something works, then it's true; if we can make apparently infallible predictions, then what we based those predictions on is true. The natural sciences have pretty much relied on that practical test of truth.

Positing "pseudoscience" in contrast to science implies that one is the opposite of the other. Yet the subjects so often called pseudoscience seem to develop in parallel with advances in mainstream science (Bauer 1986–87); "it is psychology that spawns pseudosciences and therapeutic cults" (Leahey and Leahey 1984:viii).

The attempt to distinguish science from pseudoscience is not itself a matter of natural science. Therefore we shouldn't expect to find generalizations as precise as scientific laws, or even clearly right-or-wrong answers, about whether a given claim is or isn't science. We have to settle for generalizations that are less than universal, that apply in many but not all cases of any given sort. To acquire judgment, when to believe and when not to, when to accept what one sees and when not to, one needs both good generalizations and also the benefit of experience of actual cases, especially those where people went wrong, as a guide to when exceptions to those good generalizations may be warranted.

An aim of this book is to identify useful generalizations about the seeking of new knowledge and to test them against particular instances of knowledge fights. My concern is especially with those controversies that are largely outside the mainstream of science: about psychic phenomena, UFOs, or Loch Ness monsters ("Nessies")—anomalous or anomalist or anomalistic phenomena. But within mainstream science too there are occasional controversies in which the cry "pseudoscience" is raised, say, about "cold fusion," and those deserve to be compared or contrasted with the likes of UFOs or psychic claims.

Which Knowledge? Natural Science or Social Science?

Science studies the natural world. Extrasensory perception may seem to be a claimed phenomenon of the physical world and therefore appropriately subject to the authority of science; yet actually it is about human behavior and as such a matter for psychology at least as much as for physics. Quite generally in anomalistics, at stake are matters of human behavior and human beliefs, regarding which the natural sciences have nothing much to say whereas the social sciences (and theology) do. Consider astrology: it is about how the cosmos influences our lives. Astrology doesn't argue against the laws of gravity or electromagnetism; it doesn't claim to be able to calculate orbits better than astronomy can. The appeal of astrology is its concern with human fate. That astronomy cannot make sense of astrology's procedures, that to astronomers the astrologers may be scientific illiterates, is quite beside the point. If astrology is opposed to any accepted knowledge, it is to psychological, religious, or spiritual knowledge more than to astronomical.

Again, the fascination that UFOs exert is not because they perhaps reveal ways of traveling faster than we now imagine possible but because they touch an age-old question: Are humans the only conscious, intelligent beings in the whole cosmos? Alchemy, nowadays a standard example of pseudoscience, is often referred to as though it had been merely a superseded premodern chemistry, but it was actually a whole system or worldview much like astrology or magic, in which properties of nature and of humans were intimately and meaningfully connected.

Folklore about "moon madness"—*lunacy*—would have it that humans tend to behave peculiarly when the moon is full. Anomalistics more than science would take that claim seriously. Assume that statistical evidence of such an effect is offered. It might be because the moon exerts a physical influence as it does on the tides; or, perhaps human behavior is influenced by the phases of the moon just because folklore has convinced some number of people that it is so, and they behave accordingly—in a similar fashion as Australian aborigines tend to die when they find themselves the victims of bone pointing (Spencer and Gillen 1968:534–38; Milton 1973). So claims of this sort cannot be validly dismissed by insisting that the moon could not possibly influence human behavior because it is so far away that its gravitational or other physical effect could not be discernible by individuals.

It is not always easy in natural science to prove that something is *not* so; in social science, it's even harder. Even if some number of studies show no clearly significant statistical evidence for a moon-phase influence on human behavior, such an influence might nevertheless be strongly evident with some people, just as bone pointing may effectively kill Australian aborigines but not Freudian psychoanalysts. On this point, anomalistics shares dilemmas with medicine: statistical studies showing that some new drug has no obvious side-effects might well miss its effect on, say, pregnant women, as happened with thalidomide. No study can look specifically at every possible subset of human beings that might be affected in some idiosyncratic manner.

Claims about moon-phase effects and a host of other such matters are typically dismissed by reference to what natural science supposedly knows, revealing the extent to which some people—science groupies and dedicated debunkers of heretical claims—look to natural science as the supreme arbiter of truth. This *scientistic* attitude becomes insupportable when one stops to think about it. Nevertheless, at the level of implicit belief that shapes so much individual and public action, scientism is the prevalent faith of the modern age; the manner in which we talk about "science" in everyday discourse (Bauer and Huyghe 1999) and the metaphors from science that we draw on (Bauer and Huyghe 1999; Edge 1973) illustrate that. Just consider how often we are told that *scientific* tests—not just any old tests—have proven something to be true.

A few hundred years ago, before the Scientific Revolution and the Enlightenment, questions of truth were the concern of the church. Nowadays science has superseded the church as universal knowledge authority. Science has itself become a sort of church, and scientists are in that sense also priests (Knight 1986). Science nowadays like the church in earlier centuries feels responsible for the intellectual orderliness of society. Thus pseudoscience is *heretical* belief—not merely wrong but an actual danger to the proper functioning of society and the welfare of humankind. The passion that authority always vents against heresy is directed nowadays in the name of science against pseudoscience.

So framing issues about extrasensory perception and the like as science versus pseudoscience is quite natural to our times; yet it misses the most important point, that the appeal of much so-called pseudoscience is on psychological or quasi-religious grounds and not on purely intellectual or scientific ones. Calling something pseudoscience implies that it isn't science but ought to be. Yet why should the study of psychic phenomena resemble the work of physics or chemistry?

Scientific Knowledge: Tried and True?

It isn't commonly remembered that scientific knowledge varies from highly reliable to highly unreliable. The most reliable is so in large part because it has worked well for a long time without contradiction; but such *textbook* science is a far cry from the *frontier* science that represents research in progress or the latest breakthroughs trumpeted to and by the media (Bauer 1992a).

The interesting knowledge fights are over new knowledge, frontier stuff. We don't worry much about such pseudosciences as flat-earth theory or alchemy, disproof of which by science has long been possible through convincing demonstrations performable at will. What produces tension are the issues for which science presently lacks resounding answers.

Ask, for example, "What is the real identity of Unidentified Flying Objects?" and science is said to answer, "More than 90 percent of reports have been explained as misidentifications of Venus, Earth satellites, rockets, meteors, helicopters, marsh gas, lenticular clouds, radar anomalies, and so on. In maybe 5 or 10 percent of the cases, we don't have enough information; but if we did, no doubt the same sort of explanation would apply." That's typical of the sort of answer that those who claim to speak for science like to offer: based on solid, long-attested textbook knowledge, using the most conservative extrapolation from it, in the faith that current mysteries will eventually prove to be explainable by already understood laws.

But that is not a very good answer to offer someone who sees in current mysteries a possible opportunity to discover new laws.

Ask, "How does a placebo work?" The answer you get from science will be no better than what you hear from a practitioner of alternative medicine about acupuncture or therapeutic touch; indeed, it may well seem not as good. In both instances, "explanations" are being offered for things that actually we do not understand.

Science is at its best when it is on the most certain ground. Confronted with what is not yet properly understood, those who claim to speak for science are reluctant to admit ignorance, and therefore their answers often discount or evade. So controversy ensues on the frontiers and fringes of science. It's only natural, of course, that the bitterest controversies involve the most uncertain matters. But those are also the ones that arouse the greatest interest.

Being Realistic about Science

Whether or not something is science should be judged by what science actually is, not by what it ought to be. Much of the philosophical discussion of the nature of science, however, is not descriptive but normative, addressing how science should do its work. But philosophy's criteria for good science are widely misunderstood as describing what science actually is. Consequently much popular wisdom idealizes science. Perhaps the most common illusion is that science uses a "scientific method" that guarantees objectivity (Bauer 1992a; Bauer and Huyghe 1999). Another fallacious presumption is that science now is the same as, say, a century ago, whereas in fact its practices and organization changed dramatically around the middle of the twentieth century (Bauer and Huyghe 1999; Ziman 1994)—yet the idealized contemporary public image of science is still much the same as it was a century ago.

Comparisons between anomalistics and science as it is actually practiced will show that no sharp division can be established. In this self-consciously scientific age, which is unconsciously a *scientistic* one, that could readily be interpreted as an attack on science or a denigration of science. But that is not a necessary interpretation of what follows, and it is certainly not the purpose. Rather, these comparisons illustrate the enormous obstacles that the human search for new knowledge faces. That actual science has flaws indicates not that there's something fixably wrong with science but rather that human knowledge-seeking ambitions easily outstrip humankind's knowledge-gaining capabilities.

Indeed, recognizing human limitations makes the palpable achievements of the natural sciences that much more awe-inspiring. Increasingly over the last few centuries, fallible human beings have cooperated voluntarily, internationally, in highly self-disciplined fashion, to bring humankind the most reliable knowledge it has ever possessed. To point out that science is not perfect nor all-powerful is surely not to denigrate it. We do, after all, revere many things that are not quite perfect and many people who are not quite perfect.

My ulterior motive is not to disparage science but to suggest that serious anomalistics be allowed a measure of respect, as an honest seeking of knowledge within mysteries more intractable even than those grappled with in the mainstream of natural science.

How Agreement Is Reached in Science

Doing science means seeking consensus under communally agreed criteria for judging evidence. Something that deserves to be called pseudoscience might then be an intellectual venture that does not seek consensus, or that dispenses with agreed-upon criteria for judging the evidence.

Now cryptozoology (the search for alleged creatures typically in out-of-the-way places), parapsychology, and "ufology" are often described as pseudosciences. Are those fields then unconcerned to achieve agreement, to convince others?

Of course not.

Is it then that ufology, parapsychology, and cryptozoology characteristically fail to respect the evidence, despite their best efforts?

It isn't that simple. Competent people with appropriate credentials or expertise differ over what the evidence is and what it means.

It may be that the majority of the mainstream scientific community takes one view while the majority of ufologists (for example) takes another. But there are also respected members of the mainstream community who agree more with many ufologists than with the dismissive portions of the mainstream community, just as there are ufologists who agree on a number of points more with the mainstream position than with the more unorthodox within ufology. A distinction between mainstream science and anomalistics must be looked for in finer details of how evidence is evaluated and agreement sought.

Knowledge-Filtering: A Social Demarcation Criterion

The reliability of science is owing to the effective knowledge-assessing or knowledge-filtering (Bauer 1992a) interactions that were evolved within the scientific community over the last several centuries. Working in isolation from that community might then plausibly lead to pseudoscience. Martin Gardner made this point at least implicitly when he described Velikovsky as a "hermit" scientist (Gardner 1950).

Isolation from the scientific community is not however a matter of either-or; there are various forms of it and varying degrees of it and a variety of reasons for it. There have been almost totally isolated individuals like Velikovsky or Wilhelm Reich; but also small groups in one place, as with N rays, once in the mainstream of science but no longer of it; or international "invisible colleges," as with polywater, which again drew themselves into isolation from initial communal interactions and restraints; and there have been national political units like Nazi Germany or the Soviet Union in which political action hindered interactions between that nation's scientific community and the international one.

Not only is isolation a matter of degree; even total isolation does not *guarantee* falling into error. Being a member of a mutually critical scientific community helps enormously in separating reliable from unreliable, but it may do little to spur creativity. Isolated genius may still achieve great things; the pioneer of microbiology, Anton van Leeuwenhoek, was a hermit by choice and an extraordinarily successful scientist (Ford 1982). Again, historians have noted that Isaac Newton created his mathematical physics while forcibly isolated by an outbreak of the plague.

Intellectual (Epistemic, Epistemological) Demarcation Criteria

Knowledge-filtering in science is far from arbitrarily at the whim of the community; certain intellectual criteria govern it.

Science, like other intellectual disciplines, deals in three sorts of things: facts, approaches, and explanations (or data, methods, and theories). In the normal course of events, what Kuhn (1970) called "normal" science, routine puzzles are solved and details are added to the stock of scientific knowledge without *revolutionizing*—standing on its head—any of those three things. For example, most of the work in the currently hot field of molecular biology consists of locating and identifying genes, determining their chemical structure, looking for the protein whose manufacture each gene governs and for the mechanisms that switch the gene on and off, and comparing different versions of the gene in different states of health and in different species. Unless something quite unforeseen turns up, the new data complement the existing stock without being *revolutionary,* without calling for changes in *theory;* and the *methods* being used are standard ones even as their efficacy is continually improved. There is no difficulty distinguishing *that* "normal" science from pseudoscience because the very question of pseudoscience only arises over startlingly contrary, heterodox, revolutionary claims.

In major scientific revolutions, such as relativity or quantum mechanics, only one of that triad of data, method, and theory was revolutionized:

- Planck's novel "quantum" interpretation or revolutionary theory was based solidly on widely accepted data accumulated over a long time by accredited methods. Relativity's revolutionary theory was accepted because its predictions were not only consistent with the bulk of existing data but were also able to explain new data (about an eclipse), again obtained by long-attested methods.
- When a new method or instrument is invented, its reliability is at first judged by whether its results jibe with known facts fitting the accepted theories.

- Revolutionary data are accepted if obtained by accepted methods and if they do not obviously contradict the overarching theoretical paradigm. Thus the existence of pulsars and quasars was not doubted because the methods by which they were discovered were well-established ones, and it was not obvious that existing theories could never explain their characteristics.

Innovation simultaneously in two aspects of this intellectual troika typically causes the work to be disregarded: the relevant research community doesn't know what to do about it, isn't able to *use* it. This has been called "premature science" (Stent 1972). Contemporary examples are the searches for gravity waves and for a magnetic monopole, both of which have so far yielded only a single claimed observation over the course of many years. There were novel facts and method in each case, as also with the classic example of Mendel's demonstration of numerical ratios in heredity—mathematics in biology was novel in method and so the arithmetical facts were too. Wegener's suggestion of "drifting" continents broke with prevailing theory about how the Earth's geography came about and also with accepted method or approach—he claimed as meaningful the biological and geological relationships on opposite sides of oceans. More recently Barbara McClintock's discovery of genes moving from one chromosome to another asserted novel facts for which current theory had no explanation; so it was announced in 1951, ignored for a couple of decades, and awarded a Nobel Prize in 1983. Almost contemporaneous was Peter Mitchell's work on biological energy mechanisms, breaking with theory and using a novel approach as well—ignored or ridiculed for about twenty years but awarded a Nobel Prize in 1978.

It appears, then, that working on projects that involve novelty in two of the threesome of data, method, and theory generally means having to play a lone hand, forsaking the intensely communal activity that marks the normal course of science. Mendel, Wegener, Mitchell, and McClintock were not perhaps badgered or harassed as pseudoscientists, but their search for knowledge certainly became a lonely one as their most natural potential colleagues didn't join in their work.

Going for novelty simultaneously not just in two but in all three facets of scientific knowledge—facts, method, theory—is just not done in science; but it is typical of anomalistics—see my discussions headed "Connectedness" under "Velikovsky" and "Loch Ness Monsters" in chapter 6.

That something is professedly revolutionary, at any rate unorthodox, in all three aspects of course does not necessarily mean that it is wrong. Still, the chances that it can be correct must be smaller, the odds against it greater, than if the unorthodoxy is in only one or two of the three aspects. Science, striving for reliability above all, naturally and properly expects more resounding proof, the more revolutionary the claim.

Connectedness

In a sense, intellectual and communal aspects of knowledge-filtering are both aspects of connectedness, the first on the more technical aspects and the latter over the more subjective matters that directly invoke human judgment. Anomalistics lacks connectedness, whereas modern science displays it to a remarkable extent.

Not only are scientific research projects tightly connected to the relevant special field; modern science is also strikingly coherent across its various disciplines. All natural scientists accept and draw on the same laws, facts, and methods. Biology uses chemistry or physics as called for, say, the ideal gas laws or the Periodic Table or chemical kinetics or radioactive isotopes. Geologists—perhaps in seismology—may sometimes be hard to distinguish from physicists, or at other times (say, in paleontology) from biologists, or on occasion from chemists. Interdisciplinary projects are commonplace.

What about the Social Sciences?

To what extent do the features of science just described apply to the social sciences?

There are telling differences.

Philosophy of science has paid relatively little attention specifically to social science. When it has, however, the salient questions are seen to be significantly different from those encountered with natural science. Natural science looks for cause-and-effect relationships that, once uncovered, permit accurate predictions to be made. But social science must deal with entities that are *not* linked by simple cause-and-effect relations. People and human institutions have purposes, reasons, beliefs, and desires whose connections to actions are not simply cause-and-effect; a given purpose or desire may be followed or accompanied by a variety of actions. The consequences of purposes, reasons, beliefs, or desires are *not* accurately predictable.

This lies at the root of a fundamental dilemma in social science. There are two schools of thought as to the proper aims of social science: should it strive to become "naturalistic," that is to say, like the natural sciences in looking for entities that *can* be connected via cause-and-effect laws that would make possible accurate prediction; or should it be "interpretive," respecting the autonomy of human beings as self-actuating agents and seeking to understand individual and group behavior after the fact while acknowledging the impossibility of accurately predicting that behavior? The first approach is sometimes called in social science the attempt at *Erklären*, typified by the work of Émile Durkheim; the second, whose exemplar is Max Weber, is sometimes described as the attempt at empathetic *Verstehen*. As Rosenberg (1995) cogently argues, whether they like it or not and whether they are aware of it or not, social scientists all make a choice between those two approaches, even if only implicitly, in what they do and how they do it. Moreover there is no sound way of merging the two approaches. Going the first way is being scientistic, taking predictive power as the proper mark of being "scientific"; going the second way relinquishes any proper claim to be called "scientific" if that term means, as it does in common usage, "like the natural sciences." The dilemma is illustrated by arguments over the existence of free will, or over the "mind-body" problem; that problem remains intractable in philosophy, in psychology, and in cognitive science. In religious context the analogous dilemma is that, as between materialism and transcendent belief, there seems to be no stable intermediate ground (Chambers 1953).

In any case, as a matter of experience the social sciences lack the impressive coherence that marks natural science. Whether or not the philosophical reasons just outlined explain that, actual practice makes the absence of connectedness plain enough. Even individual social sciences lack the unity that characterizes individual natural sciences. All chemists, say, agree on all the fundamental and significant issues about chemistry: what data are sound, which methods are trustworthy, which theories are useful. By contrast, all psychologists do *not* agree on all those matters; if some agree over data and even over methods, they are still likely to disagree over what theory best explains the facts.

Social science does, of course, *seek* agreement just as natural science does. But since social science is "multiparadigmatic"—each discipline encompassing separate and competing theoretical paradigms—it rarely attains and does not really expect universal agreement over interpretation of data or even over what the best or proper methodological approach is. In the natural sciences, disputes over novel claims may concern new methods, or new data, or new theories; in the social sciences, disagreements tend to be more between different schools of thought.

Given that the natural and the social sciences display these salient differences, it will often be necessary in what follows to consider separately how anomalistics compares with natural and with social science.

Is Difference Pejorative?

Identifying differences is an intellectual exercise; it isn't automatically an evaluation of relative merit, worth, or importance. Still, the delineation of differences between the natural and the social sciences seems often to be taken as somehow pejorative toward the social sciences. Pointing to a lack of coherence or consensus is easily seen as belittling, especially where matters of knowledge are concerned.

Now it is true that some physical scientists have regarded social science as inherently inferior, as when a mathematician campaigned against admitting to the National Academy of Sciences a social scientist whom he called a practitioner of pseudoscience (Sherman 1987). But merely *questioning* the scientific status of the social sciences—wanting to be clear about it, in other words—is not the same as denigrating those intellectual disciplines; unless, of course, one insists that "science" *à la* natural science is the only respectable form of intellectual discipline. Such prominent social scientists as Stanislav Andreski (1972) and Ernest Gellner (1984) have themselves queried the scientific status of the social sciences (Bauer 1992a:128–40; Turner and Turner 1990); according to the eminent anthropologist Clifford Geertz (Horgan 1996:156), "nothing in anthropology has anything like the status of the harder parts of the hard sciences, *and I don't think it ever will.* . . . Things get more and more complicated, but they don't converge to a single point. They spread out and disperse in a very complex way. So I don't see everything heading toward some grand integration. I see it as much more pluralistic and differentiated" (emphasis added).

In my own view, "science" is *not* the only worthwhile human intellectual activity, nor the only proper or possible source of knowledge, nor the only arbiter of what correct knowledge is. Social science is valuable, as are the humanities, quite irrespective of how they compare with natural science. Much of my own writing (Bauer 1984a, 1986; Martin 1988) is more "social science" or "humanities" than it is natural science or "science." For me, recognizing differences between natural and social science is not inevitably a condemnation of either one. Indeed I've argued that while natural science is preeminently reliable on matters that are physically demonstrable, those matters carry little if any meaning for human existence (Bauer 1995b; Bauer and Huyghe 1999). It is the humanities and the social sciences that deal with questions of human existence and human meaning. James Q. Wilson (1997) has put the essence of it nicely:

> Real science is most likely to be produced (although not always) when scholars study inanimate objects, as do biologists and chemists. The science that seeks to explain human behavior . . . is much less empirically exact. At best, it consists of

> little more than an assertion that, in a certain percentage of cases, a person with certain traits (such as a particular score on the Minnesota Multiphasic Personality Inventory) is likely to be part of a group that will be more likely than others to behave in a certain way.

That Wilson puts biologists among those dealing with "inanimate objects" illustrates that there is here, as so often, no truly unequivocal distinction to be made. While such biological subdisciplines as biochemistry do study matters that are physical more than behavioral in nature, one surely cannot exclude from "biology" the study of animal behavior. Still, the distinction between natural science and social science is clear enough for the present purpose: between, respectively, certain and merely probable consequences of a given set of circumstances. That's the essence of it, and for many purposes it is a world of difference.

What Gets Called Pseudoscience

The "Big Three" subjects in anomalistics are: parapsychology, ufology, cryptozoology. But a veritable host of other matters are often lumped together with those (Dotto 1978; Fraknoi 1984; Frazier 1976, 1978; Kraatz 1958): ancient astronauts, anti-fluoridation, anti-vaccination, anti-vivisection, astral projection, astrology, Bermuda Triangle, biorhythms, chiropractic, Christian Science, creationism, dianetics, immortality, Kirlian photography, lost continents, moon madness, near-death experiences, orgone energy, plant communication, pyramid power, the Shroud of Turin, Sirius B and the Dogon (The Sirius "mystery"), Velikovsky's claims, von Däniken's claims . . .

The excellent annotated and analytical compendium by Leonard George (1995) has several hundred entries, almost all of them related to psychic or religious phenomena. Cavendish (1989) has a comparable number of entries plus a similar number of synonyms and cross-references. Jerome Clark's judicious and well-documented *Encyclopedia of Strange and Unexplained Physical Phenomena* (1993) has about 150 main entries and an index of 1,500 topics, people, and publications, with almost no overlap with psychic topics—Clark emphasizes cryptozoology, ufology, and such oddities as crop circles and reported alligators in sewers. The most comprehensive and up-to-date collection of reported anomalies of nature is owing to William Corliss (1994), who has culled from scientific periodicals some 40,000 items ranging chiefly over the natural sciences.

The examples cited above show that it is not easy to discern a substantive or topical basis for lumping these subjects together. What does the Bermuda Triangle have to do with Kirlian photography, or either of them with what the

Dogon believe about Sirius B? Is there in fact any family resemblance among all these subjects, other than that there is argument over their validity? What exactly is it about them that causes the charge of pseudoscience to be leveled?

In seeking similarities it is natural to consider into which of the accepted intellectual fields of inquiry the subject might fall. But in anomalistics typically it is not easy to specify exactly what the subject matter is. Is extrasensory perception a psychological phenomenon? Or a spiritual one? Or a physiological or anatomical one, analogous to seeing by means of eyes or hearing by means of ears, possibly by means of an as-yet-undiscovered organ?

Thus anomalistics has some of the characteristics that all interdisciplinary searches for knowledge have, or searches for knowledge in fields that do not yet belong to any recognized discipline. However, interdisciplinary projects in science have usually called on closely related disciplines: hard sciences and engineering in building the atomic bomb, say. It is worth noting that anomalistics attempts the further stretch across both natural and social sciences.

We cannot group things logically together until we know exactly what they are. Thus Cavendish (1989) cross-references "auras" and "Kirlian photography," whereas I would place "auras" under "mind-mind interactions" and Kirlian photography under "electrical-photographic measurements." If, however, auras turn out to be objective phenomena then Cavendish could be right and I wrong.

In this difficulty of classification, anomalistics resembles natural history, the precursor to modern science. In one fifteenth-century compendium, there was a grouping based on the shapes of objects, hardly an illogical criterion to employ at a time when little more was known about the objects. However, among the things thereby classified together by conical shape were belemnites, which are of mineral composition; fossil sharks' teeth; and prehistoric axe- or arrowheads (Ashworth 1980). It required a great subsequent expansion of scientific knowledge over several centuries and in several fields before a more meaningful, functional classification became possible.

Since anomalistics by definition deals with ill-understood matters, the relationships among the various topics are usually uncertain. And it is not only that clearly anomalous topics are not readily classified nor easily distinguished from one another: in a number of instances there is also no sharp division between the "normal" (or the mainstream) and the "anomalous." Cavendish (1989) has entries for Freud and Jung, yet those figures and their theories are at least as much part of the historical mainstream of Western culture as they are anomalous.

Many of Cavendish's (1989) entries are for individuals; and this brings out another reason why anomalies often resist classification. Restless intellects like

Carl Jung, Wilhelm Reich, and Rudolf Steiner had notions that not only ranged from mainstream to idiosyncratic but were also not restricted to propositions about a single class of phenomenon.

The tendency for anomalistics to seek legitimation through association with respected subjects also leads to a blurring of categories, as does the attempt by some anomalists (often to the dismay of other anomalists) to link together various unorthodox claims more or less by brute force, as in the occasional suggestion that some cryptids (the animals cryptozoology searches for) are so elusive because they are psychic manifestations (Beckjord 1988; Holiday 1986).

What to Call These Subjects?

The difficulty of categorizing these subjects is matched by the lack of a satisfactory global term for them. What I am calling anomalistics, what many people call "pseudoscience," has been given a variety of other names: Forteana, enigmas, oddities, "alternative science," popular fallacies. All have one or another drawback if one wants to approach these matters in neutral fashion.

The inveterate nay-sayers use not only "pseudoscience" but also "occult," "superstition," "paranormal," "Cargo Cult science." Emblematic of this "science-police" is the Committee for the Scientific Investigation of Claims of the Paranormal (CSICOP), which publishes *Skeptical Inquirer* and is self-consciously dedicated to fighting pseudoscience in the public media, for example, the astrological horoscopes carried in the newspapers. CSICOP-type debunkers are also referred to by their opponents as sci-cops, CSI-cops, or CSICOPpers.

The "skeptical" in *Skeptical Inquirer* and the "skeptics" in the names of many groups employing that label interprets skepticism in the sense of those ancient Greeks who actively *dis*believed, the *a*theists, rather than in the nowadays more commonly understood sense of agnostics, people who suspend judgment, who maintain an attitude of doubt. CSICOP and its "Skeptics" are doubtful only about unorthodox beliefs, which they judge in the light of the contemporary scientific knowledge that they do *not* doubt. Those who doubt what the "skeptics" believe, for example concerning the Holocaust (Shermer 1997), are called "fringe," "pseudoscience," and the like by these self-styled "skeptics." In point of literal fact, anomalists could call themselves "skeptics" with just as much warrant as those who debunk pseudoscience, since anomalists are genuinely skeptical in relation to much of currently accepted scientific theory (though not usually of scientific knowledge, which is by no means to be equated with scientific theory [Bauer and Huyghe 1999]).

Do Names Matter?

Not only "skeptics" but many other groups and things as well assume or are given names that may be somewhat inappropriate or misleading. In science, once-fitting names can become unfitting through the advance of understanding. Electrons, photons, and the like were named when they were believed to be particles. Now we no longer believe them to be particles, but the names have stuck. Outsiders, people unfamiliar with science, those who rummage in scientific works looking for support for their ideas without properly understanding modern science, can be led astray by such names that have become misleading, as when Immanuel Velikovsky in 1950 compared the solar system with the "electrons" "orbiting" the nucleus of an atom (Bauer 1984a).

So it does matter how one refers to the subjects I call anomalistics: calling them "oddities" might also be neutral enough not to be a prejudgment, but talking of them as "fallacies," "superstition," or "pseudoscience" implies that instant dismissal is warranted, that taking an interest would be foolish, that one already *knows* they are spurious.

Degrees of Implausibility

Since anomalistics embraces topics that are somehow apart from what we take to be normal, one way to classify them might be by the degree of their differentiation from the normal. Arthur C. Clarke (Welfare and Fairley 1980:8–11) suggests mysteries of the First, Second, Third, and Zeroth Kinds. The First Kind used to be mysteries until science solved them: rainbows, for example, before it was known that white light is composed of all the colors and can be separated into them by prisms or raindrops. The Second Kind mysteries are like UFOs or sea serpents, where quite plausible reasons for existence are available, if only we knew for sure that the phenomenon exists and what exactly it is; given enough time, these will undoubtedly become First Kind, with explanations that may be anywhere from disappointingly banal to wonderfully surprising. Third Kind mysteries are like spontaneous human combustion or poltergeists, which seem entirely inconceivable. Even if such incredible phenomena demonstrably occur, there would seem to be no way of using that knowledge because no known mechanisms exist—they would at best be just like the *premature discoveries* that science ignores. Clarke's Zeroth Kind mysteries are ones that ought never to have been granted the status of mystery at all, like the Bermuda Triangle, where there never really was enough evidence that anything anomalous has gone on.

A somewhat similar distinction is proposed by Marcello Truzzi, between *crypto* and *para* claims. Crypto claims (compare Second Kind mysteries) assert that certain objects are real—sea serpents, say. Those could quite simply be proved beyond doubt: a carcass would do it. Once the existence of such a cryptid is indubitable, science has no difficulty in incorporating the knowledge. (But crypto claims often cannot be *dis*proved, certainly not in any simple fashion. "Absence of evidence is not evidence of absence," is a cryptozoological aphorism.) Para claims (compare Third Kind mysteries), by contrast, assert that relationships exist that contradict our current views, for example that bodies can be made to move by sheer mind-power in absence of a physically measurable force.

Significant vs. Ignorable Claims

Anomalistics covers an enormous range of subjects; but not all of these inspire serious investigation as opposed to attention by entertainment or news media. Some topics, Clarke's Zeroth Kind, can be dismissed without much ado, for example most of what the *National Inquirer* prints, or assertions that the Earth is hollow and its interior inhabited by some sort of intelligent beings. Such things can be dismissed forthwith because no shred of even potentially credible evidence is offered for them. This book is concerned not with those but with what serious, genuinely curious, and by-and-large competent investigators of anomalies are interested in: the seeking of knowledge about what seem to be mysteries, the satisfying of curiosity about claims that, though implausible, seem not entirely impossible.

Of course and as always, no sharp line can be drawn, in this case between the serious and the ignorable claims. Judgments of plausibility vary. Competent opinion will often differ over whether a purported phenomenon is to be taken seriously or not. Still, in most cases the contrast is clear enough: it is between, on the serious side, the assertion that here are mysteries to be solved and, on the other hand, blandly dogmatic assertions of "truths" that contradict established scientific knowledge. The serious anomalist asserts that acupuncture, therapeutic touch, and the like are worthy of study because they are practiced and believed and therefore present at the very least a social conundrum that needs to be understood—and because just possibly though not necessarily they may also point to some physical or medically useful practice; whereas a spurious, ignorable claim would be the assertion that Krebiozen or laetrile or sitting in pyramids or orgone boxes are already known to be better than conventional treatments for cancer and should be used and paid for in preference to conventional treatments.

2 Anomalistics and Science Compared

Consider some commonly asked or argued questions:

1. What exactly are the energy-producing reactions inside stars?
2. How can we best help children to high achievement in school?
3. Can some people sometimes get accurate information through a "sixth sense"?

It may not seem immediately obvious that these questions are of drastically different kinds. In fact they are; these questions belong to disparate realms of intellectual discourse. The first is a question for natural science, the second for social science, the third for anomalistics.

Perhaps the salient difference is in the possibility of delivering an assured yes-or-no answer. No one doubts the value of research on the first question or the feasibility of arriving at a definite answer. Regarding the second, there is a certain division of opinion about how to get an answer and how reliable any answer could be, as circumstances vary; but few would argue that one shouldn't *try* to find the best possible answer. Those who seek answers to these first two questions are not told that they are doing something silly. What makes the third question so different from the other two that loud voices persistently call it a misguided question, an instance of gullible superstition, an example of pseudoscience?

To understand that means comparing anomalistics with the natural and the social sciences. How best to make that comparison? Surely by looking at the actual practices of seeking knowledge in those realms. But there is no generally accepted description of what those practices are. What follows is therefore inevitably and heavily influenced by personal experience. The reader is then owed some insight into the author's relevant experience:

- Natural Science: My education led to degrees in chemistry; from the 1950s I was a university teacher of chemistry and a researcher in electrochemistry. I read regularly the leading general periodicals in science as a whole, *Nature* and *Science* in particular. Three of my books are on chemical topics (Breyer and Bauer 1963; Bauer 1972; Bauer, Christian, and O'Reilly 1978).
- Social Science: My curiosity about science's attitude toward unorthodoxies prompted me to change my professional field from electrochemistry to science studies (also often called STS, for Science and Technology Studies), a mix of the history, philosophy, and sociology of science, technology, and medicine—perhaps a "humanities" discipline as well as a social science, but primarily the latter and certainly not a natural science. At least one of my books (Bauer 1992a) and possibly another (Martin 1988) is indubitably social science; another two (see the next paragraph) are either social science or anomalistics.
- Anomalistics: In the early 1960s, I became fascinated by the possibility that Loch Ness monsters (Nessies) might be real animals. By the 1970s, I had become fairly certain that they are, and therefore I became curious why mainstream science ignores them. That led to interest in other anomalist subjects. Since 1980 I've been active in the Society for Scientific Exploration, which was formed specifically to provide a scientific forum for considering such matters. For some twenty years I've also read much in anomalistics, in the primary literature of proponents and critics as well as the relatively scanty analytical literature. I've published case studies of the Velikovsky affair (Bauer 1984a) and the Loch Ness controversy (Bauer 1986).

So I have been active in the three fields that I seek to compare. Besides publishing books and articles in all three, I attended professional meetings, gave seminars, and served in organizational roles. Some early insights into differences between the natural and the social sciences I also gained from eight years as dean of a college of arts and sciences (Martin 1988).

Why do the three questions posed at the beginning of this chapter belong to what seem like different realms of intellectual discourse? Because the character of any field of intellectual inquiry is shaped by the questions that field seeks to answer. Anomalistics is concerned with mysteries, and much of its character flows inevitably from that. It makes little sense to ask that parapsychology be "scientific" if that means applying methods suited to long-established fields with copious amounts of long-attested data, since parapsychology lacks methods that are already known to produce reproducible data. But does such lack make it silly to try to find out whether humans do occasionally display a sixth sense?

Tables 1 and 2, below, list some broad-brush comparisons between natural science, social science, and anomalistics that are discussed at somewhat greater length in the text.

Contrasting Natural Science and Social Science with Anomalistics

> *Much has already been achieved in the established disciplines; about anomalistic subjects, little is yet known.* A great deal flows from this obvious point.

Established disciplines became established by making substantial progress in answering questions they are concerned with. They evolved approaches that proved effective in expanding knowledge and therefore became consensually accepted.

Anomalistics by contrast knows only which approaches have so far been *ineffective.* Parapsychology has tried and abandoned mediums and seances and card-guessing experiments, for example, which were in vogue for long periods. Nessie hunters have learned that keeping watch doesn't yield many sightings, that fish bait and hormones don't attract Nessies, that sonar gives lots of echoes that cannot be satisfactorily interpreted. Investigators favor a variety of approaches because none has been obviously effective.

Table 1 contains points on which anomalistics is contrasted with the established disciplines in both natural and social science.

As part of each discipline's progress, agreement ensues as to what constitutes satisfactory proof of a given point:

> *The established disciplines have their own standards of "proof"; anomalistics does not.*

Natural science requires a law to have no exceptions. In social science, the standard is lower: "proven" means convincing, but within each of the social sciences, there are several schools of interpretive thought—in economics, for example, a range from Marxist through libertarian. What's convincing to one may not be to the others. On quantitative matters, however, there is wide consensus on the commonly used "95 percent probability level" of statistical

Table 1. Contrasting Natural Science and Social Science with Anomalistics

Natural Science	Social Science (and Humanities)	Anomalistics
much has already been achieved		*little* is yet known
there is *consensus* over what is worth doing		there is *no consensus* over what is worth doing
"proven" means *no exceptions*	"proven" means *convincing,* but there may be opposing schools of thought. In quantitative studies, a common criterion is the "95% probability level": *wrong 1 in 20 times*	there is no accepted standard of proof. The ambition is to have the (natural) sciences accept the anomalous claim, which entails a high level of reproducibility or overwhelming statistical odds
the relevant discipline is *obvious*		the relevant discipline is *not obvious*
much "normal" activity: building on what's known		there is *no* "normal" activity
there is *high probability of useful results*		there is *low probability of useful results*
progress is *steady with occasional leaps forward*	(the sense of progress is not present in social science as it is in natural science)	progress is *all or nothing*—typically nothing with very occasional successes
anomalies are ignored (until the next scientific revolution comes along)	(what role is played by anomalies in social science?)	*anomalies are the real stuff*
research focuses on the *known unknown*		research focuses on the *unknown unknown*
luck favors *the prepared mind*	(what role is played by luck in social science?)	what would be the best preparation?
common sense is *beside the point*		common sense is *the chief guide*
practitioners *exceed some minimal level of competence*		anyone can play; *wide range* of competence *and incompetence*
hoaxes and fraud are *relatively rare*		hoaxes and fraud are *very common*
the *high points* are emphasized by the media		the *low points* are featured by the media
controversies are *handled internally* (except when social applications are at issue)		*controversies* are *in the public arena*
sponsors give but researchers decide what to do		sponsors and patrons tend to attach strings

significance. This means that "facts" that are wrong once every twenty times are accepted as valid.

In anomalistics, "proof" means acceptance by the established disciplines.

What discipline is appropriate?

In the natural and social sciences, there is a subspecialty ready to tackle any question that gets asked. If need be, several subspecialties cooperate and sometimes new specialties emerge thereby. In the early twentieth century, for example, biochemistry was developed to answer questions about the chemicals found in living things.

In anomalistics there isn't even the illusion of being clear what disciplines are relevant. Ufologists may need to call on aeronautical engineering, meteorology, astronomy, assessing of eyewitness reports, analysis of folklore, use of hypnosis to uncover repressed memories, and more.

Progress?

Science progresses steadily during long periods of accumulating not particularly startling knowledge—"normal science"—interspersed by occasional great leaps—"scientific revolutions" (Kuhn 1970). In anomalistics, by contrast, the accumulation of information doesn't mean, as it does in the sciences, a steady expansion of *useful* knowledge. Occasionally there may be apparent advances, as when underwater photos of Nessies were obtained in the 1970s, but after a time it becomes apparent that little if any true progress has been made. We seem to be no closer to knowing what UFOs are than fifty years ago, no closer to identifying Nessies than in the 1930s, no closer to understanding psychic phenomena than in the late nineteenth century.

In mainstream academic fields, it makes sense to tell new investigators to choose some topic of research that is almost guaranteed to produce publishable results within a few years. It would be fatuous to tender that advice to one who wants to practice cryptozoology, parapsychology, or ufology.

But those "useful results" in the sciences might also be described as the steady accumulation of relative trifles; major advances are rare. Anomalistics eschews the scientific work of steady building upon what is already known and instead seeks the occasional major jackpot.

But has anomalistics really won any major jackpots?

Yes. It is undeniable that some formerly anomalistic or heretical subjects have later become accredited by mainstream science, either as true or at least as worthy of serious investigation. Classic cases include meteorites (Westrum 1978) and the mythical kraken recognized as the actual giant squid (Heuvelmans 1968:45–79). The first (dead) specimen of the extraordinary platypus sent to Europe had been regarded as a fake. More recent times have added acupuncture, archaeoastronomy, astronautics including artificial Earth-satellites, and the search for extraterrestrial intelligence (SETI); ball lightning (Barry 1980); continental drift; the timing of the first human arrival in the Americas; the language represented by the Linear-B script and that on the Phaistos Disk.

Those are not the only ones, and surely they are not the last heretical claims to be vindicated. Among scientifically rejected but cryptozoologically sought creatures there are such relatively plausible possibilities as the Tasmanian tiger and the South American ground sloth as well as the more improbable yetis (ape-men of the Himalayas) and Mokele-mbembes (living dinosaurs in the Congo).

There is an ironical aspect to anomalist success. If the unorthodox claim turns out to have indisputable substance, then anomalistics loses its monopoly on the subject and science takes over. Physicists and meteorologists now study ball lightning; giant squid are grist for marine biologists; meteorites have long been part of geochemistry and planetary science; and so on. Anomalistics proposes but science disposes.

So what distinguishes a *scientific* discovery of an unexpected animal from a *cryptozoological* one? It is that zoology looks for things to do that are as immediately productive as possible, whereas cryptozoology is willing to take long chances based on folklore and dubious tales and mythical-sounding descriptions by pre-scientific peoples. In science, such tales and reported inexplicable sightings are ignored as "just anecdotes."

Anomalies are ignored in science; they are everything in anomalistics.

Natural science ignores "outlier" results unless similar ones occur over and over again. Kuhn's (1970) view is that scientific revolutions eventuate when ignored anomalies have so accumulated that they become too prominent for the conventional wisdom to disregard any longer.

I am not aware of any similar discussion of the occurrence or role of anomalies in the social sciences. Given that the "mainstream" in the social sciences typically harbors several competing views—Keynesian, free-market, public choice, Marxist, and others in economics, for example, and given the different standard of proof in social science, whereby up to 5 percent outliers are statistically acceptable, it may be that nothing in social science is seen as anomalous in the fashion that Bigfoot is or ball lightning was seen as anomalous in natural science.

For anomalistics, the real grist is the things that science ignores or cannot explain, what might be called the "unexplained residuum," such as all the UFO sightings that haven't been explained as Venus, satellites, and so on. On this criterion, "scientific creationism" is quite clearly not science but anomalistics: the main thrust of its arguments focuses on those things that theories of evolution by mutation, natural selection, neutral drift, and other known processes have not yet explained, such as macro evolution and the complexity and integration of living systems (Behe 1996; Johnson 1991; Macbeth 1971).

But not only anomalistics regards the unusual as significant. The practice of medicine, for example, cannot but be concerned with outliers and rarities. Infrequent individual effects are regarded as important, certainly not as spurious: that one person in ten million gets polio from the vaccine, or that some people with very high cholesterol don't get heart disease, or that some people infected with HIV never progress to disease. It may be that in anomalistics, analogous rarities could turn out to be the whole story, say, if psychic ability is something that only a few people have to any marked degree, or only at particular ages (adolescents attracting poltergeists, for example), or if psychic events happen only under very unusual circumstances (telepathy or clairvoyance when someone is dying, perhaps).

In mainstream natural science, attention to the unexplained anomaly has characterized some major breakthroughs in science. Well known is Alexander Fleming's refusal to ignore the petri dish in which for some reason his bacterial culture didn't grow—just as well, for he discovered penicillin thereby. Szent-Györgyi (1971:2) attributes all of his original contributions to paying attention to some small detail overlooked or ignored by others: "the discovery of ascorbic acid was due to the observation of a small delay in the reaction of peroxidase with benzidine, a reaction done thousands of times daily. It is done even in elementary courses of biochemistry. The discovery of actin and actomyosin rested on the observation that after storage the stickiness of my muscle extracts increased. When I showed this to H. H. Weber, the leader of muscle research, he said that he has seen this many times but thought that his preparation went wrong and sent it down the sink. That I did not do likewise may have been due to the many wild theories I made which prepared my mind."

But most work in science follows the conventional wisdom, and most scientists focus on the standard rather than the outlier results. A somewhat similar way of putting this point is that anomalistics tries to search in the *unknown unknown* whereas the established disciplines venture cautiously into the *known unknown.*

What's *known* thereby defines some things as waiting to be learned: knowing the structure of DNA, we realize that all sorts of genes and interactions await discovery. Doing research toward that end is studying the known unknown, doing Kuhn's "normal" science. The periodic scientific revolutions bring entirely unanticipated and unforeseeable things; they pop out of the unknown

unknown: around 1900, radioactivity, quantum mechanics, relativity; in the 1960s, compounds of inert gases; in the 1990s, bucky-balls (pure carbon arranged in symmetrical cages of sixty atoms, in hexagons and pentagons rather like miniature soccer balls or Buckminster Fuller's geodesic domes).

Mainstream disciplines behave as though the unknown unknown doesn't exist; perhaps just because it cannot be directly investigated. Anomalistics carries on as though it knows pretty much what is ready to pop out of the unknown unknown.

Luck.

It is generally recognized (at least by historians and sociologists) to what high degree success in science depends on luck (Hannan, Roy, and Christman 1988; Hetherington 1990; Stephan and Levin 1992). Watson and Crick could not have succeeded at a different time or in a different environment. Jaroslav Heyrovsky won a Nobel Prize for following up a phenomenon whose very existence depends on the presence, then unsuspected, of impurities in smaller amounts than could at the time be detected (Jain et al. 1979). But it is also true that, in science, luck favors the prepared mind. It usually takes a very knowledgeable, competent, alert scientist to take advantage of what luck offers.

While the importance of serendipity in natural science is generally recognized, I am not aware of similar discussions of it in social science. One instance is Derek Price's realization (1986:xix) that the growth of scientific activity has been exponential. He had stacked successions of annual volumes of certain journals on the floor and it struck him how the tops of the stacks curved upward.

Luck plays an even greater and much more random role in anomalistics than in science. Since the phenomena are not well enough defined to be studied in a controlled way, one has to be lucky to get any positive results at all, be it in attempting to observe UFOs, telepathy, or Bigfoot. Since it isn't clear what methods might work, or which disciplines might be relevant, one cannot even deliberately prepare oneself to be ready to take advantage of it if luck happens to strike.

It might seem to contravene common sense to attempt research with as little guidance on how to go about it and as little hope of success as there is in anomalistics. But common sense by no means dictates human actions, and many people now remembered as great were not praised for common sense in their own time, for instance, Winston Churchill in the decade before 1939.

Common sense is deployed in anomalistics but not in the sciences.

Considering not the common sense of engaging or not engaging in an anomalistics quest, but the deployment of common sense during an actual investigation, one comes on a paradox. In modern science it is actually viewed as being beside the point.

In modern natural science, the deep, sophisticated understanding that has been won is quite often at variance with everyday common sense. Quantum mechanics insists on such ludicrous things as "tunneling," for example, whereby a particle that hasn't enough energy to surmount some barrier nevertheless has a finite chance of reaching the other side. (That's explained as happening because these "particles" are really waves as well as particles, or actually neither but behaving sometimes as one and sometimes as the other.) The only excuse science can offer for such absurdities is that, when used strictly within their particular range of application, they correctly predict tangible, observable consequences.

The attitude of social science toward common sense is an even more curious one. It seems to assume that it can establish expertise only if, as in the natural sciences, it is able to command a body of understanding that the laity cannot share because it runs counter to common sense: "what sociologists say about common sense is the self-serving ideology of a vested interest group seeking to establish and maintain a monopoly over 'its' professional turf" (Pease 1981:266). "'Common sense' has received widespread negative treatment in introductory sociology textbooks" (Mathisen 1989:307); "we must . . . let go of a 'common sense' theory of learning," theorists of education insist (Wildman 1996:2). Economics Nobelist James Meade (quoted by Wolpert 1993:6) suggested for his epitaph, "He tried to understand economics all his life, but common sense kept getting in the way."

Lacking the depth and sophistication that natural science has attained and that social science strives for, anomalistics habitually appeals to everyday common sense. These appeals then often seem insubstantial to those schooled in science. It makes excellent everyday common sense to believe eyewitness testimony and personal experiences of psychic phenomena sufficient to establish to the satisfaction of just about any court of law that the testimony is trustworthy; but psychologists and others who know about the vagaries of human perception find that argument less than compelling—even leaving aside that courts do not aim to establish *scientific* truth.

The established disciplines train new practitioners; anomalistics is free for all.

That much has been achieved in the sciences entails that neophytes must undergo disciplined training before they can become accredited contributors to further research; but there exists no such training for people who want to study UFOs or Nessies. Nevertheless, some anomalist researchers are as competent as any in the mainstream of science; many, however, are not, and some are more incompetent than all but a very few in the established disciplines.

The personalities of researchers in mainstream science and in anomalistics seem to be more overlapping and similar than their technical competence tends to be. One finds in both groups a wide range of human characters. One aspect of human behavior is however far more frequent in anomalistics than in natural or in social science—fraud, and the related behavior of hoaxing.

Fraud is rare in the sciences but common in anomalistics.

The temptation to deceive others, be it for tangible gain (fraud) or fun (hoax), is surely succumbed to more often when the deception is least likely to be discovered. The solidity and cohesion of much of science makes it quite difficult to avoid being found out in relatively short time (Bauer 1992a:42–62). In active fields of research, hoaxes or frauds are quickly detected and eliminated: within a year or so, judging from the periodic reports from the National Institutes of Health (1997) or the notorious modern cases summarized by Broad and Wade (1982) in *Betrayers of the Truth: Fraud and Deceit in the Halls of Science.*

In fields where it is difficult to acquire data, though, deceptive claims may cause trouble for a longer period of time; the Piltdown hoax was a conundrum for paleoanthropology for many decades, and even thereafter came decades of argument over who the perpetrator(s) might have been (Walsh 1996). Prehistory and archaeology offer a number of such instances—see, for example, *Great Exploration Hoaxes* (Roberts 1982) and *Fantastic Archaeology* (Williams 1991). There continues to be controversy over artifacts like the Holly Oak pendant (Kraft and Custer 1985; Sturtevant and Meltzer 1985) or the Kensington runestone. For many examples of hoaxes and references to further reading, the *Encyclopedia of Hoaxes* (Stein 1993) is very useful.

In social science, perhaps because almost any set of data can be accommodated by some respectable theoretical approach or other, error introduced by hoax may persist for a long time, for example over the sexual behavior of Samoans (Freeman 1983).

In anomalistics, where facts and theories and methods are all up for grabs, hoaxes can be readily devised and may be detected only by chance or only after a long time.

Hoaxes or frauds in science are typically viewed as aberrations. In anomalistics, on the other hand, any hoax or fraud is suggested by debunkers to demonstrate that the whole subject is spurious or fraudulent.

Hoaxes and frauds represent the extreme of controversial claims that need to be settled. The long-established disciplines have evolved ways of handling technical disputes; anomalistics lacks these.

The media feature the accomplishments of the sciences; the "news value" of anomalistics lies in its absurdities.

Because of its tangible successes and consequent prestige, science tends to be judged by its high points and its heroes; only in recent decades has there been some muck-raking reportage of science, in books that describe the careerism of scientists "warts and all" (Bauer 1992a:166) and in bringing to public attention instances of scientists found guilty of scientific misconduct. Anomalistics tends to be judged in the other extreme, by its low points. In innumerable popular and semipopular articles, serious and trashy anomalistics are lumped together. For the media, "newsworthy" are the great charlatans, the flawed cult-leaders, the errors.

Established disciplines carry on and also settle their own arguments internally; anomalistic disputes are in the public arena.

In funding research as well as in settling disputes, anomalistics lacks the autonomy that the established disciplines enjoy. Research typically costs something; one needs sponsors or patrons.

The established disciplines decide internally what to do with the funds they are provided; patrons of anomalistics tend to attach strings to their gifts.

Modern science has long been freed from patrons who told it what to do with the resources provided. That's been a traditional distinction between basic *science* and applied *technology*, between seeking knowledge and applying it. Until quite recently, science and the wider society both agreed that if, say, high-energy particle physicists said they needed a Superconducting Super-Collider, they should have it, cost be damned. Patrons of science have long come to accept that even as they pay the pipers, the pipers themselves will call the tunes. It's virtually taboo within scientific circles to suggest that the choice of research projects should be made other than by the group of specialists in each field and subspecialty.

Research in anomalistics suffers from lack of resources available for investigators to do what they know to be required, as opposed to what those with money to give would like to have done. While there are a few—a very few—foundations and individuals that seek or purport to behave rather like the National Science Foundation in supporting good work in anomalistics in response to proposals from researchers, more common by far are strong-willed patrons who try to force research into procrustean directions.

Scientists suffer conflicts of interest between seeking knowledge, building a career, being loyal to an employer, and satisfying sponsors. Anomalists suffer those in somewhat different forms, and often a particularly distinctive one as well. Anomalistics being a hobby and not a profession in most cases, personal interest in an anomalist field may be unwelcome to one's professional associates and employers. Scientists are usually well advised not to take a public interest in anomalies in their own field. As a chemist, I suffered little if any professional disdain because I made no secret of my fascination with Loch Ness monsters; but had I been a biologist, that fascination could have had unfortunate professional corollaries, as a number of actual examples illustrate (Bauer 1986:123–24).

Contrasting Natural Science with Social Science and Anomalistics

On a number of points (see table 2), social science occupies somewhat intermediate ground as between natural science and anomalist research, for example:

> *The technical literature in natural science is tightly organized, in social science less so, in anomalistics not at all.*

The coherence of scientific knowledge is matched by the coherence of its literature. Finding what's known on a given subject is made easy by compendia of data; by indexes by author, keywords, and so on; by periodicals that publish abstracts of research articles; and by citation indexes that lead one to any article that has cited a particular piece of work. There is also a long-established tradition of popular yet trustworthy exposition of science in books and periodicals.

Much of this apparatus also exists in the humanities and the social sciences, though there are instructive differences. There are few if any parallels to the review journals of the natural sciences, understandably so, since the latter's purpose is to monitor progress, which is not the salient feature of humanities or social science that it is of natural science. Speculative or interpretive or theoretical articles are more common outside the natural sciences. Compendia of data are not as prized as in the natural sciences, although some census data and the like are quite often used in social science, and concordances and bibliographies are standard resources in the humanities. Citation indexing came later to social science and is not yet part of humanistic studies.

Anomalistics lacks any such organized literature. It features anecdotes, forays, expeditions, *attempts* that may not have met with any substantive success, what would in science be regarded as therefore not publishable. Neither enthusiasts nor debunkers are much interested in attempts at unbiased evaluations of the state of the field. Most anomalist claims are reported in the popular media, which are exceedingly unreliable on these matters (MacDougall

Table 2. Contrasting Natural Science with Social Science and Anomalistics

Natural Science	Social Science (and Humanities)	Anomalistics
highly organized literature; little falls through cracks	*fairly well organized* literature	*unorganized* literature; *recurrences*
universally shared paradigms	*competing* paradigms or interpretations	
experts are *reliable guides*	experts are *reliable* guides (*if you accept their ideology*)	experts are *not reliable* guides
get there first (if one doesn't do it, someone else will)	*exemplify* an interesting approach or interpretation that may be unique	*define a whole field*
strong conservative quality control	judgment of quality is influencedby *ideological preferences*	*no* quality control
evidence must be *objective:* anecdotes and eyewitnesses don't count	*testimony* of trained "participant observers" *admissible*	much of the evidence is *personal testimony from untrained observers*
replicability is *essential*	replicability *cannot always be demanded*	replicability is *rare,* until mystery is solved
keep novelty within bounds	if method is sound, novel data are accepted	no bar on novelty: the more, the better
stable organizations	occasional schisms	perpetual schisms
society accepts internal legitimation	society accepts internal legitimation (but ignores some schools of thought)	legitimation must come from outside, from "science"
nothing is supernatural or nonmaterially based	body-mind problem unresolved	cannot exclude supernatural

1983), or in cult periodicals, usually founded by one particular group of enthusiasts and often short-lived. Monographs are the most common vehicle for publication of an overview of anomalist subjects, but almost always they present just a single viewpoint and do scant justice to others. Much of this material is self-published. Proper textbooks, manuals, or histories rarely exist, though some attempts have been made in those directions in parapsychology in particular (Beloff 1993; Berger and Berger 1991; Broughton 1992; Coly and McMahon 1993a,b; Wiseman and Morris 1995). Compendia of data are not often available, even when they would be highly desirable, as for instance comprehensive listings of reported sightings of Nessies. There are, however, some catalogs of UFO sightings and UFO encyclopedias (Clark 1998). Abstracts and citation indexes don't exist.

Refereeing poses particular problems in anomalistics. In science, referees can judge a manuscript's worth by assessing its method, theory, and facts in the light of canonical knowledge. Referees and prospective authors think along the same lines about the criteria for publication, and the rate of rejection of manuscripts submitted for publication isn't large, except in the most prestigious outlets, such as the journals *Nature* and *Science.* In the humanities and the social sciences, by contrast, where there aren't the same agreed criteria of method, facts, and theory, rejection rates can be as high as 50 to 80 percent or even 90 percent (Franks 1981:117). Ideological divisions result in somewhat less consistency of refereeing than in the natural sciences (Broad and Wade 1982:102–3; Coughlin 1988; Peters and Ceci 1980), and some periodicals consciously cultivate a particular ideological approach. In anomalistics a few such journals as *Cryptozoology, Journal of Scientific Exploration,* and *Journal of UFO Studies* attain commendable standards through strong editorial control coupled with conscientious refereeing; nevertheless a great deal of rubbish is disseminated about anomalist subjects, especially in newspapers, general magazines, and nonrefereed specialty journals.

In science, the popular media serve only the purpose of popularization. In anomalistics, however, these are the very sources of much of the data; investigators of UFOs and the like often learn initially from the mass media of the events they want to study. Charles Fort's pioneering compendia (1974) of unexplained oddities drew on what had been published in newspapers and magazines.

Book reviews are a very useful guide to a given body of literature. In science, such reviews tend to be as consistent as the refereeing of journal articles, influenced once in a while by personal disagreements but rarely by ideological ones (except on such topics as evolution, where atheism or Marxism wield much influence). In social science, reviews of a given book quite often are polarized along ideological lines into strongly favorable and strongly unfavorable; see, for example, reviews of *Higher Superstition* (Gross and Levitt 1994), which held up to ridicule some postmodernist attacks on science. Knowing only the subject of a book and its author's viewpoint, one can often predict successfully, in the humanities and social sciences but not in the natural sciences, what the tenor of a review is going to be; say, with Shelby Steele's collection of essays, *The Content of Our Character* (Steele 1990), or "in sociology, [where] the state of book reviewing is, if anything, far more ideologically motivated, far more superficial, and, as a result, far less responsible in its treatment of academic books and their authors" (Wolfe 1998:B4). So too in anomalistics: reviewers are likely to have their pet commitments, and conflicts with that are liable to produce a virulent reaction. Thus in a favorable review of my book on the Velikovsky affair, Martin Gardner voiced pained incredulity that I could believe Loch Ness monsters to be real (Gardner 1985); in a favorable review of my book about the Loch Ness controversy, the usually judicious Jerome Clark was moved to anger because I appeared to dismiss evidence for the reality of UFOs (Clark 1987); and while that book reaped high praise from almost all quarters for its objectivity in the face of my personal beliefs, including from such unbiased sources as *American Rationalist* (Stein 1987), a reviewer for *Skeptical Inquirer* found it possible to deplore my bias (Kelly 1987).

When a scientific discovery is made, any earlier and previously ignored work is quickly rediscovered, as with Mendel's laws of genetics. When claims were made in 1996 of fossil life in a meteorite, it took little time to ferret out that similar claims had been made in the 1960s (Urey 1966) and the 1880s (Birgham 1881). But when a *spurious* discovery is announced, it is less easy to uncover if it had already been found wanting in the past. There exists no comprehensive account of all the premature or false trails that science has taken. By and large the history of science has focused on the successes of science. So it is understandable that the same false trails may be taken at intervals, and it seems characteristic of anomalistics that there is a periodic recurrence of the same or similar heresies and arguments about them:

- The biorhythms fad of the 1970s was a rerun of Wilhelm Fliess's ideas of half a century earlier.
- Physical scientists were fooled by sleight-of-hand of mediums in the late nineteenth century, and again in the 1970s and more recent years.
- To demonstrate that psychic phenomena are spurious, the stage magician Houdini offered a monetary reward to any claimed medium or psychic whose abilities Houdini could not mimic. Half a century later the stage magician James Randi offers the same challenge.
- Perpetual-motion machines of any given type are periodically reinvented (Ord-Hume 1977:55).
- Kabbalistic Gnosticism—using numerology to unearth secret truths in ancient texts—has "been resurrected over and over again, persisting in one form or another for over two thousand years" (Aveni 1996:64); it came up again in 1997 with claims that certain combinations of letters in the Hebrew Bible could not possibly have arisen by chance (Drosnin 1997).
- Palmistry "peaked in popularity in the sixteenth and seventeenth centuries . . . [and then] lay submerged for nearly two hundred years before . . . resurfacing in the New Age" (Aveni 1996:220).

- Astrologers like to mention as legitimation of their field that it has a very long tradition, whereas historical investigation reveals that astrology had died out in Western civilization until revived in quite recent times.

Such recurrence and repeated failure to become accredited does not, however, mark a claim as definitely spurious. Many claims of high-temperature superconductors had been wrong, but eventually some have proven correct. That a succession of discoveries of the "first planet outside the solar system" (Croswell 1997; Flam 1992b; Glanz 1997) turned out to be erroneous doesn't mean that we will never actually find one; perhaps the claims of the last few years will finally stand up.

The highly organized literature of science reflects the high level of integration of the natural sciences:

Natural science is monolithic; social science and anomalistics encompass disparate schools of thought.

In the natural sciences, disagreements at the research frontier tend to be over detail, over specific and particular matters, while the broader body of facts, methods, and interpretations is communally agreed, which provides a framework for settling the disputes over details.

In social science, disagreement is often about overarching issues of interpretation. Incommensurable, competing schools of thought differ over how to interpret human behavior, ranging widely from various sorts of determinism and behaviorism to quasi-mysticism like that of Carl Jung. As to political behavior, Marxism competes with a variety of other interpretive stances. As to overall worldview, the social sciences harbor various shades of realism and relativism.

In anomalistics, interpretive stances range from highly skeptical all the way to various degrees or kinds of belief. For example, everyone who takes an interest in UFOs believes they are "real" in the sense that witnesses are not lying (though nearly everyone also concedes that what is reported may not be what actually happened). But explanations run the gamut. There is the extraterrestrial hypothesis—UFOs as alien spacecraft. Various psychosocial hypotheses seek explanations in terms of culturally shaped misperceptions, hallucinations, or visions. There have been tentative speculations about time-travelers. In cold-war times there were those who suggested that UFOs were secret enemy vehicles. Nowadays there are still those who believe them to be secret American military gadgets. A useful introduction to the vast literature on UFOs is Clark's encyclopedia (1998); another explanation mentioned there is the "psychoterrestrial" one: UFOs are "forces of rebirth"; "mythic constructs . . . [that] obligingly affect radar" (706).

Consonant with these differences as to paradigms, experts play different roles in natural science, social science, and anomalistics:

> *In natural science, the experts are reliable guides; in social science, they are reliable if you share their ideology; in anomalistics there are no reliable expert guides.*

In the natural sciences, when experts tell you how to do something they are usually right about it. In the social sciences, recommendations will come as much or more from ideological commitment as from sheer technical expertise, and predictions are not to be heavily relied upon—for example about how to raise educational achievement or get people off welfare. In anomalistics, experts only command knowledge of the state of the field and its history but cannot give assured guidance to further investigation. Experts on Bigfoot can tell you about claimed evidence, who the researchers are, what the media coverage has been, and so forth; they do *not* know what Bigfoot are or how to study them effectively. So these are not experts in the usual sense: expert carpenters, expert mathematicians, expert physicists can all do something reliably and on demand; experts on anomalies cannot.

In established disciplines the best work is usually done by the experts. In anomalistics, any breakthroughs are so likely to come by pure chance that it is often a neophyte rather than an expert who makes it: the best film of a Loch Ness monster was obtained by Tim Dinsdale on his very first expedition, and though he became increasingly an expert, he didn't get another photo or film in more than twenty-five years—while some other people did who just happened to be in the right place at the right time.

> *Fame in natural science comes to those who make a discovery first; in social science to creators of a new approach; in anomalistics to those who define a whole field.*

Most historians of science agree that the great discoveries were bound to be made by someone, sooner or later. Evidence for that lies in the frequency with which the same discovery has been made simultaneously and independently by two or more people (Merton 1957, 1961; Price 1986:59–61), for instance, conservation of energy or the planet Neptune.

In the social sciences, there is no agreement over whether a given approach would ever have been taken up were it not for the particular individuals who did so; would psychoanalysis have eventuated without Freud? In anomalistics, whole fields are owing entirely to a single individual: Wilhelm Reich and orgonomy, say, or Ron Hubbard and dianetics and Scientology.

The mistaken belief that scientific progress can come relatively easily and quickly, at least for great scientists, may be a contributing reason why some anomalist gurus are able to attract followings. According to the popular view, such geniuses as Newton, Darwin, Einstein, Watson and Crick, and numerous others were able to push science forward stunningly more or less alone; so why shouldn't there arise other gurus able to do the same? Charismatic people like Immanuel Velikovsky and Wilhelm Reich may seem to the public via the media to be quite plausible polymathic scientists, capable of building by themselves in a single lifetime a whole edifice of valid albeit alternative science. Were it more widely understood that the reliability of science depends on the knowledge-filtering functions of scientific *communities* (Bauer 1992a), then it might become less common that single unorthodox gurus can entice large followings while standing aloof from and in contradiction to mainstream science.

The jockeying for prominence in science is well disciplined by accepted norms of behavior and criteria for preferment. By and large, one needs the approval of peers. In anomalistics, jockeying for position often is less a matter of seeking the approval of peers or making contributions to the field than of attracting attention from the media. Anomalistics therefore makes news more through the charlatans, hoaxers, and absurdities that plague it than for its serious investigations.

Judgments of quality show an understandable variation from science to anomalistics.

In each discipline practitioners learn what the standards of competence and quality are, and these do vary in different fields. The more that's known on a subject, the more accurately can work be judged. In social science it's sometimes difficult to separate ideological differences from judgments of quality; a behavioral psychologist may not be a good judge of the quality of Freudian analysts, for example. In anomalistics, so little is known that judgments of what work is of high quality are more subjective—though it remains, as in all fields, rather easy to recognize *poor* quality.

The requirement that evidence be objective and that it be replicable shows a similar variation.

Psychology has learned about the variety of ways in which we fool ourselves through misinterpretation of sensory perceptions, of other people's motives, of the probabilities of coincidences, and so on (Gilovich 1991). A personal experience, no matter how subjectively convincing it may be, is—speaking objectively or scientifically—a very poor reason for belief. So eyewitness testimony proves little if anything in science—just in a few pockets like field biology.

In some social sciences, personal testimony is accepted as sound, for example, from psychotherapists or from anthropologists doing field studies.

In everyday life, however, personal testimony can be decisive, as in courts of law; and anomalistics tends to follow this everyday common-sense approach. In anomalistics, direct personal experience is often what creates true believers. Of course, individual scientists doing mainstream research also often become true believers in their own discoveries. But in science, unusual claims are *communally* accepted only after being attested by sundry people in different laboratories. Anomalistics lacks the communal control that the sciences enjoy, and so it is replete with true believers relying on their personal conviction.

Personal experiences are not repeatable on demand; and being able to repeat observations is often taken as a hallmark of science. But social science cannot demand replicability since it deals with animate self-willed beings. In anomalistics, replicability doesn't exist; if their facts were reproducible, cryptozoology would be zoology and parapsychology would be psychology.

Novel data are more easily accepted in social science than in natural science; in anomalistics, only novel data are of interest.

Contrary to the popular view that science is always seeking novelties, natural science is actually an extremely conservative pursuit. It resists novelties that contradict what it already believes to know. Even eventually successful revolutionary novelties typically encounter strong resistance before they become generally accepted (Barber 1961; Estling 1982).

Social science is otherwise. Any data obtained by accredited methods tend to be accepted, and there are few if any "anomalies" that cannot be incorporated into one or another of the competing schools of interpretation. But in some fields that overlap the natural and social sciences, or if there is not a preexisting congenial school of thought, resistance to novelty can be ironclad. In American archaeology, for example, until very recently the mainstream dogma has been that the Clovis people, characterized by the earliest then discovered occupation site, were the very first to populate the New World. Tentative clues to earlier human presence were pushed aside. Yet it was always quite unlikely, surely, that the earliest so far discovered site should coincide neatly with the earliest actual human presence. Again, in the decipherment of ancient scripts, extreme conservatism seems to have retarded progress on more than one occasion: in the breaking of the Maya code, the dogma that the script was purely pictographic and without phonetic representations (Coe 1992); with Minoan Linear A and the Phaistos Disk, that pre-Mycenaean culture was not proto-Greek—a prejudice perhaps even more remarkable because in the 1950s, and against considerable resistance, the decipherment of Linear B had already overturned the "Establishment" view that *it* couldn't be a Greek language (Estling 1982; Fischer 1997).

In anomalistics, which is betting on something revolutionary turning up, it often seems as though the more novel a claim is, the more favorable attention it gets. As someone once remarked, whereas for conventional disciplines mysteries represent something of a failure, for the true anomalist believers mysteries may be a triumph.

In anomalistics, resistance to novelty is naturally low since the aim is to break into the unknown unknown. There is no reliable body of accepted relevant knowledge against which to gauge the validity of new theory or data or method. Anomalistics is *exploration,* where everything may turn out to be different than on home territory; so open-mindedness is practiced to a degree that may seem excessive to those used to scientific practice. The tendency is to believe a claim until it is disproven, whereas in science one disbelieves or suspends judgment until a new claim is proven. Admittedly, when challenged, anomalists will allow that the burden of proof is always on a new claim to prove itself; but anomalistics deals intrinsically with implausible phenomena for which the evidence is not yet compelling, so in practice anomalist research must proceed as though it had already satisfied the burden of proof.

Established fields have organizational as well as intellectual stability; anomalistics has neither.

The sciences enjoy well-established and typically stable organizations—professional associations, journals, accrediting bodies, although in social science, ideological divisions can interfere.

The characteristics of a community are not a simple sum of the native individual attributes of its members; each community develops practices and mores appropriate to its particular purposes. Many of the detailed differences between science, social science, and anomalistics are associated with differences in the maturity, sophistication, and international character of their organizations. Those organizational differences then amplify characteristics of the fields' practices. Thus much of the strength of science stems directly from the efficient, workmanlike, task-oriented procedures of the scientific community; and the weaknesses of anomalistics have much to do with the lack of such communally governed practices.

However, meetings and journals of anomalist organizations are not readily distinguishable, on external grounds, from mainstream academic gatherings and periodicals. The ability of commercial hotels to make all the necessary arrangements for professional meetings, and cheap computerized typesetting and desktop publishing, combine to allow all sorts of groups and gatherings to give every superficial appearance of professional depth even when sound intellectual substance may be largely or wholly lacking.

The natural sciences have been the focus of work by international communities for several centuries. The coherence of scientific knowledge is paralleled by the accord and rapport that scientists feel across national and religious divides. Atheist, Buddhist, Muslim, Christian, Jewish; Marxist and capitalist; Russian, American, African, and European scientists work amicably and productively together. "Scientist" is a term with which the manifold specialists in the natural sciences are comfortable: they share the same axioms in method and theory and the same conservative demand for intellectual rigor.

There's nothing forced or artificial about this cohesion of the scientific community. It exists because of evolutionary developments over the course of centuries. Geologists and physicists and other natural scientists recognize their mutual kinship because of the manifest similarity and compatibility of their interests and approaches and the way in which their detailed facts and methods and theories overlap and interlock with one another. In science, mainstream disputes don't produce permanent schisms in the research community; differences of opinion tend not to have major organizational consequences. Occasionally but rarely do research institutions create separate units just because some people disagree so fundamentally, even though engaged in the same investigation, that they can no longer work effectively together, but when this happens it is commonly as the result of clashes of personality or differences over ethical behavior rather than over ideological or substantive scientific issues.

By contrast to the natural sciences, the social sciences don't have nonideological international communities. The natural cohesion one finds among natural scientists is not displayed among social scientists or humanists. "Social science" is a term with which many in the specialized fields are comfortable, but to a somewhat lesser degree than people in the natural sciences are with "science"; for example, historians are split between those who accept it and those who insist they are humanists and not social scientists. Sometimes the various social sciences may seek to work jointly on matters of clear common interest, say, in seeking society's moral and tangible support; but that's different from the instinctive compatibility that unites those who work at the disparate natural sciences.

The social sciences do have coherent national communities, but within the same topical field there are internal divisions. The American Psychological Association split, for example, over the manner in which basic science and clinical practice do or should connect (Coughlin 1995; Holden 1994b). Freudian psychoanalysts convened in 1998 "to repair the damages caused by innumerable organizational splits—beginning with Adler and Jung over 85 years ago" (Kurzweil 1998). Anthropology suffers tension between those leaning toward natural science and those leaning toward social science (Selvin 1991; Shea 1998).

Communal cohesion within both natural and social sciences is encouraged by the fact that community membership offers practical benefits: professional status, group insurance, help with gaining and keeping employment and with further education. Intellectual curiosity may be the prime and original reason why such a community comes into being, but thereafter the stability of that community is further cemented by economic and social benefits.

Anomalistics is even less a matter of natural compatibilities than is social science, and membership in anomalist groups promises solely intellectual benefits. Much of the time that may be enough, but when opinions and attitudes differ about the topic of interest, groups easily splinter. Schisms are typical and common rather than infrequent. There are several neo-Velikovskian groups and a multitude of orgonomic organizations—"science cults never die, they just slowly fade after the death of their charismatic gurus" (Gardner 1988:28). There are a plethora of UFO organizations—to the outsider, for no good reason; yet the insiders are so convinced that their own approach is right and the others' wrong that they choose to proceed separately. So anomalist organizations tend to be small, to split whenever any significant internal disagreement crops up, to come into being and disappear again rather quickly. These internal disagreements are not often brought to public attention by the regular media, but they are well known to insiders and are reflected in a variety of bulletins, newsletters, open letters, and so on; a fine illustrative source for gossip about disagreements among individuals and groups in ufology is James Moseley's newsletter *Saucer Smear,* now available on the Internet.

Debunking "skeptics" groups, too, have experienced significant schisms. Soon after CSICOP began publishing its journal, its editor (and cochairman of CSICOP) resigned in view of a fundamental difference: "Paul Kurtz . . . considers that belief in the paranormal constitutes a threat to science, whereas [Marcello] Truzzi sees it as a threat . . . to conventional religion. . . . There is . . . a spectrum of opinion on the committee between those who tend to favor a harder-line, debunking treatment of the paranormal and those who tend toward a skeptical but open-minded assessment of paranormal claims" (Wade 1977). Another crisis for CSICOP came when some of the committee "bungled their major investigation, falsified the results, covered up their errors and gave the boot to a colleague who threatened to tell the truth" (Rawlins 1981:1). In Germany the Forum Parawissenschaften (1999) was organized in reaction to the dogmatism of the "skeptical" debunking group.

Within the various anomalistic fields one sometimes sees attempts at an appearance of solidarity in the face of the dismissiveness and contempt displayed by science, media, and "skeptics." The clearest instances of this are the typical refusal of anomalists to discuss their differences publicly or to admit, as they privately believe, that some of their number are incompetent or worse. As with other groups, anomalists are afraid that, should they admit corruption or incompetence on the part of anyone who is regarded as one of their number, the wider society would take all of them to be corrupt or incompetent—especially perhaps because the usual level of respect for anomalistics is not high in the first place. Of course this is misguided and self-defeating in the longer run, but it's typical of all guilds and groups. Even mainstream science resisted for a decade or two before acknowledging that there has been in the second half of the twentieth century an increasing incidence of fraudulent conduct by scientists in "hot" fields like molecular biology and clinical medicine (Broad and Wade 1982).

A few organizations seeking to cover particular anomalistic fields have attained a measure of stability and respect: the International Society of Cryptozoology (ISC), the Parapsychological Association, the Center for UFO Studies. But even the ISC had to suspend publication for several years—these associations depend heavily on volunteer effort and generous patrons. It may well be that most serious, active anomalists associate themselves with such organizations, but still their membership is tiny in comparison to the number of people in fan clubs and the like, who provide much of the audience for books on anomalistic subjects.

Anomalistics as an overall realm of knowledge seeking has even less intellectual coherence than the individual anomalist topics do. "Anomalistics" is a term that few ufologists, parapsychologists, cryptozoologists, or others interested in unorthodoxies would even recognize, let alone happily accept as describing them (though I trust they will prefer it to the most common alternative, "pseudoscience"). Bigfoot enthusiasts and those who hunt dinosaurs in the Congo may respect one another when they stop to take notice, but they rarely communicate with one another and have little natural or instinctive affinity with one another. There is no general feeling of commonality between ufologists and cryptozoologists, or between either of those and parapsychologists. It's only recently that some few have recognized the value of an association like the Society for Scientific Exploration (n.d.), which offers a forum for anomalist discussion irrespective of topic and including the sociology and history of anomalies. In a sense, "anomalistics" may be to individual anomalist topics a little like "philosophy of science" is to individual topics in the sciences: only a small proportion of actual researchers is interested in these broader implications.

Maverick scientists who find mainstream colleagues unwilling to take seriously their unorthodox views sometimes welcome the opportunities and audiences afforded by anomalist organizations. Thus the prehistorian Euan Mackie found in Velikovskian circles an appreciative audience for his unorthodox interpretations of megalithic society. Professional astronomers whose views are not taken up by the mainstream, such as Victor Clube and Tom Van Flandern, are featured speakers at neo-Velikovskian conferences. The Society for Scientific Exploration has been an appreciated venue for discussion of such unorthodoxies within mainstream science as cold fusion, quantized redshifts, Thomas Gold's notions of where oil deposits can be found, and Benveniste's experiments with ultra-dilute immunogens.

Such associations may offer moral support, but they also present insidious intellectual hazards. The maverick mainstreamer comes to be on familiar terms with anomalists; comes to respect some of them; comes to understand their viewpoint. From understanding viewpoints it's not a long step to considering them seriously, and then seeing them as less implausible. So some mavericks are led further and further into anomalistics, to views that are increasingly beyond the pale, from mainstream views on water electrolysis to cold fusion to alchemy and unorthodox energy theories, for instance (addressed in such periodicals as *New Energy News* and *Fusion Facts*); or from crop circles as possibly a meteorological phenomenon, to their role in relation to sacred sites like Stonehenge (Vidal 1991) or mythical ones like Atlantis (Meaden 1991, 1997).

Society respects established disciplines; anomalistics is given no respect.

The wider society allows considerable self-governance and self-determination to scientific organizations, including on such matters as the distribution of research funds provided by the public or the government. Surely the organizational stability and record of substantive achievement and effective internal quality control in mainstream fields has led to this trusting delegation. Thus science has effective external relations; the wider society almost invariably accepts the scientific community's internal judgment—even when that is counterintuitive, as with relativity or quantum mechanics. By contrast, anomalistics is saddled with a spotty if not entirely tarnished image. Anomalistics has no effective internal legitimation, so society usually asks established science for its judgment over anomalous claims.

On some occasions, however, popular opinion does find expert scientific judgment unacceptable: for instance, the majority of people believe that something like psychic powers do exist, despite the fact that no mainstream discipline does. In these cases, much of the mass media falls in line with public opinion and is more open to the anomalist side: in the Velikovsky affair (Bauer 1984a), respected mainstream pundits among humanists and social scientists long supported the anomalist position against the scientific one.

> *Science has no room for the supernatural; anomalistics and social science cannot dismiss it* a priori.

In fundamental presumptions that are more implicit than explicitly propounded, there is a significant difference among anomalistics, natural science, and social science. Natural science has progressed by presuming that all its questions can be answered without recourse to supernatural explanations or investigations. Social science, dealing with human behavior, can hardly exclude a priori the existence of intangible human senses or powers, at least before the nature of the relationship between mind and brain has been elucidated. Anomalistics even more cannot exclude supernatural considerations.

Summing Up

There are, then, these broad differences between anomalist pursuits and research in the sciences. But there are also considerable variations among and within the natural sciences, and among and within the social sciences, as well as within anomalistics. Some subfields in those three realms therefore bear greater similarity to one another than these broad-brush contrasts might indicate. For example, cryptozoology, heavily reliant on eyewitness testimony, bears similarities to "field biology," the study of how animals behave in their natural environment; one who spends months apart from other human beings but in the company of gorillas or chimpanzees has experiences akin to some of those who seek dinosaurs in the Congo, and may find similar difficulty in having some those experiences believed.

Cold fusion, polywater, and N rays were all generated smack in the middle of competent, genuine mainstream science, yet pundits have not hesitated to label them "pseudo" or "pathological" science—even though that epithet is not applied to *all* cases of mainstream science gone in some manner wrong. Various aspects of the relations between electromagnetic forces and life processes have over the last few centuries been sometimes part of mainstream science and at other times part of the denigrated "pseudo" realm. Such cases, considered in chapter 5, will serve to illustrate further how particular topics are not readily placed in a sharp category of "pseudoscience" or anomalistics or natural science or social science.

Broadly speaking, anomalistics ought perhaps to be compared not with "science" but with natural history, what science was before it had accumulated much reliable technical knowledge; or with frontier science, the unreliable cutting-edge stuff that typically needs much redrafting and rethinking before *some* of it turns out to be valid.

But there are significant differences too between anomalist research and frontier science. The initial results from mainstream scientific research indeed should not be readily believed, and much in them will later be replaced; but that work proceeds in a larger context of connections to well-established relevant facts, methods, theories. The judgment of possible validity is helped and constrained by those connections. Anomalist research by contrast lacks useful connections to well-established facts or methods or theories; it is often almost sui generis. As indicated above, that has profound effects on the judgment of quality and of validity and of competence, on the usefulness of the literature, and on the stability and influence of organizations. There are also consequent characteristic differences in the controversies one encounters in mainstream fields and over anomalistics.

3 Knowledge Fights

By its nature, anomalistics is controversial. It seeks knowledge where the conventional wisdom holds there's none to be found: "All the territory we cover is disputed" (Cavendish 1989:7). The history of an anomalistic subject is at the same time a tale of controversy.

Now some things are common to almost all disputes, and they feature in anomalistics just as they do in arguments of other sorts. We close ranks to defend our bedfellows, almost irrespective of how incongruous or self-defeating that may become. Arguments feature obfuscation and red herrings more than attempts to settle the issue by evidence and logic. Mountains are absurdly made out of molehills like spelling and punctuation (Bauer 1984a, e.g., 62, 241).

Substantive issues get mixed up with personal ones. It tends to become a matter of *who* is right rather than of *what* is right. People on each side come to identify themselves with the issue, it becomes a cause. Winning is what counts. Things are taken personally, arguments are made ad hominem, as though if the opponent can be discredited, the substantive argument has been won. One seeks ways of saving face, of putting it as though one had not been wrong.

The protagonists are often trying not so much to convince their opponents as to impress some other audience. The arguers often don't mean what they say; even when they intend to, often they're unable to say what they really mean; and in no case should their statements be taken at face value.

Bystanders may presume that some commentators are neutral, disinterested pundits, but often there are hidden agendas in play. In the Velikovsky affair, certain social scientists who claimed to be concerned with modes of procedure and not with the rightness of Velikovsky's claims turned out to have more reasons than that to take his side (Bauer 1984a,b).

Odium scholasticum is commonly encountered. The quarrel is more passionate between those whose views might seem to outsiders to be closer to one another, than between those who are farther apart—schismatics and heretics arouse more holy fury than do heathens or barbarians. People who share closely similar views can become outraged that the similarity is not identity.

The committed believers and the committed disbelievers are often not poles apart, at least not in personality—they may in fact be carbon copies with respect to dogmatism, finickiness, and the like. If they change, typically they don't just drop the error of their ways but go to the opposite extreme, committed Communists becoming the most outspoken anti-Communists, such as Arthur Koestler or Whittaker Chambers. Here are some examples in anomalistics:

- The most persistent critic of neo-Velikovskiana is Leroy Ellenberger, who had once been captivated by Velikovsky.
- The psychologist Daryl Bem changed from disbeliever of psychic phenomena to believer in the reality of remote viewing.
- Debunkers of Loch Ness monsters include a number of erstwhile believers who became disillusioned, among them Maurice Burton (Burton 1961), Steuart Campbell (Campbell 1991), and Ronald Binns (Binns 1984).

Such switches may bolster the fallacious notion that there are only two sides to the issue and only one of two possible answers. But in anomalistics there are usually many possibilities (see "That One Side Is Wrong Doesn't Mean the Other Is Right," in chapter 7).

In science—in academe generally—it is extraordinarily rare for people who disagree over intellectual questions to seek satisfaction through taking their case to a court of law. But anomalistics lacks the institutions and behavioral norms that regulate controversy in mainstream science and scholarship, so it's far from uncommon to find arguments sprinkled with threats of legal action and occasionally actual law suits—among Reichians, for example. Uri Geller sued CSICOP and Randi. And the sci-cops too sometimes threaten to sue when their intellectual discourse is criticized. For example, in a generally favorable book-review a reviewer ventured that "I feel that his own [referring to the book's author] investigation is a useful object lesson for the reader in the dangers of selecting evidence which supports what appears to be a personal bias. . . . I think the author unwittingly illustrates his own admonition about the need to be vigilant against personal prejudice and wishful thinking" (De Maigret 1994:293–94). Outraged that his objectivity be questioned, the author threatened to sue the journal that had published the review.

Presumptions

Disagreements often hinge more on unstated fundamental beliefs or attitudes than on the concrete manifest issues that the argument is overtly about.

We Believe Because It's True

"How could anyone believe that?" is our common reaction when confronted by the indubitable fact that some people *do* believe what we ourselves do not, say, that there exist ghosts, witches, yetis, or living dinosaurs. The underlying presumption is that everyone ought to have the same beliefs because we believe—or should believe—only things that are true.

But of course that's far from the case. Human beings do believe the greatest variety of things, often mutually contradictory things, often with little if any evidential support.

Many people tend to believe whatever they're told—even by con men. Others tend to believe the opposite of whatever they're told. Few indeed are skeptical and empirical in disciplined fashion. The real mystery about belief is not how we come to believe something, but rather how some of us are able sometimes to change our minds under the force of evidence and logic rather than of emotion.

Still, the passion in many arguments testifies that the opposing sides imagine or presume that somehow the others should agree and could be made to agree. Certainly they speak and act as though they believed that. Perhaps it's an inevitable corollary of a human wish for certainty.

Science Has the Answer

Present-day culture presumes that science is the ultimate arbiter of truth (Bauer and Huyghe 1999), but different people apply this variously in any given instance. Debunkers insist that in any dispute, contemporary science should win. Science itself tends to ignore anomalies or to set them aside in the confidence that an explanation will come in due course. Anomalists tend to agree that science is in principle authoritative but holds that contemporary "Establishment" science happens to be wrong on their particular pet issue; what science cannot now explain they call unexplain*able* by science rather than simply as yet unexplained.

Debunkers point to the contradiction of the conventional scientific wisdom, which is textbook science. But anomalistics is less comparable to textbook than to frontier science. Many or most of the marvelous scientific achievements also stood in contradiction to textbook science when first proposed, and were ignored, resisted, and even reviled. What once was called pseudoscience occasionally turns into real science. Anomalists are often familiar with this, hence their frequent comparison of their own situation with that of Galileo versus the church.

To be intellectually sound, criticism of an unorthodox claim cannot prove incorrectness simply by pointing to its contradiction of established knowledge; it must show why this particular unorthodoxy is unlikely to triumph in the future as however did those of Arrhenius, McClintock, Mendel, Wegener, and many others, indeed including Galileo.

Facts or Theories?

"Extraordinary claims demand extraordinary proofs," is a common aphorism. But fundamentally the issue is, rather, whether to trust empirical evidence or contemporary scientific theory. The opposing sides usually fail to recognize how close this lies to the root of their polarization.

Controversies in mainstream frontier science also involve typically unidentified opposite reliance on "facts" and on theories. An instructive contemporary illustration is with respect to global warming (Cairns 1997); the empirical data do not point to anything indubitably alarming, whereas some theories and especially some computer-generated models do.

In the natural sciences, the ultimate test of truth is empirical: nature decides what is and what is not the case. But it is only in the long run that scientific theories come to be increasingly in tune with the way the world is experienced; and therefore in well-developed and mature areas of science, claimed empirical contradictions of established theory and practice are often ignored in the short run—since observations or experiments may well be in error, such apparently contradictory experience is often ignored. Thus claims of exceptions to Faraday's Law, a law that had served well for a century, were ignored by most electrochemists in the confidence that those experiments must have been flawed—which was indeed shown some years later (Kónya 1979).

In anomalistics, the true believers tend to pose as determinedly empirical: these anomalous events, they say, simply happen (or simply have happened). Eyewitness reports, folklore, tidbits from old documents, and the like are all accepted as at least provisional data. The debunkers, on the other hand, stand on the existing theoretical paradigm: such things didn't or don't happen because they cannot. The philosopher David Hume is constantly cited as to the possible occurrence of miracles: it is always more likely that human beings are lying or mistaken in their perceptions, than that anything should happen that our scientific laws hold to be impossible.

But current scientific knowledge is not necessarily the last word. The possibility of at least apparent miracles cannot be denied a priori if only because, as one of Arthur C. Clarke's laws notes, "Any sufficiently advanced technology is indistinguishable from magic."

Impossible: There's No Way It Could Happen

It seems natural to reject reports of some happening when there's no plausibly conceivable mechanism by which it could occur; in other words, when there's no available "explanation" of how it could occur. Regarding, for example, "a study of whether 'intercessory' prayer—prayer for someone else's recovery—works . . . , a significant effect . . . [would] 'necessitate taking the next step and asking what is the mechanism by which that effect works. [With prayer], we can't even begin to know how to ask that'" (Roush 1997:357).

But if one could show a correlation between prayer and improved recovery, should it really matter that the mechanism remains unknown? Are there not many things that we accept to happen though we don't understand how they do, such as psychosomatic illness or the placebo effect (Harrington 1997; Shapiro and Shapiro 1997)? To explain those we would need to understand how mind and brain interact, which is universally acknowledged to be a prime mystery; yet the healing and the illness do both indubitably happen at times.

The implacable demand for "mechanism" reveals a strict materialism. Those who insist on it are not really relying on "science" so much as asserting that nothing in life can have transcendent or supernatural causes. But for a religious believer this is not the case, and prayer itself *is* the mechanism by which intercessory prayer works.

The dilemma of facts that lack explanations is inescapable in clinical medicine, alternative medicine, psychology, and parapsychology. "'We are vitally interested in mechanism. But there is another fundamental issue here. . . . Mind may never be defined in reductionistic terms, because of its very nature.' . . . [I]f researchers can't say how the brain gives rise to consciousness, they can't ascertain exactly how mental states initiate changes in the body. 'Therefore, to be criticized for studying only the effects of psychological states is an absurdity'" (Roush 1997:358).

Even aside from the mind-brain or mind-body problem, anomalistics is replete with instances of reports being dismissed because they can't be explained by current scientific theories:

- No physical object could accomplish the maneuvers claimed of some UFOs.
- If someone could look into the future, actions could be taken to make the future different from that peek into it, thereby producing a logical contradiction.

- Benveniste's observations of allergenic activity in solutions of homeopathic-level dilutions: there aren't enough molecules of the active reagents present to produce the claimed effect (Davenas et al. 1988; Maddox et al. 1988).

The trouble with that sort of reasoning is that even some purely material phenomena are indubitably real despite our inability to explain them. Cosmic rays are generated by phenomena whose energy is of a magnitude that baffles our ability to conceive a mechanism. The homing instincts and communicating abilities of insects are unquestioned, while our explanations for them are tentative at best. The Ice Ages did occur, but we don't understand how or why they came about. And so on.

In the past, some of the most excellent arguments later proved to be false, as to why something just could not be so. That meteorites could have fallen from the sky was absurd, for where would they have come from? Everyone knew that there are not stones hanging up there in the sky. Wegener's notion that the continents had drifted apart was also absurd, for what mechanism could possibly generate the required energy? Arrhenius's suggestion that salts exist in water as positively and negatively charged species was absurd, since what could keep those from attracting each other?

Those all seem fine arguments. It's just that they were incorrect, as in many other cases of resistance by mainstream science to the startlingly new (Barber 1961). Stones *do* hang up there in the sky, though usually far away in the solar system and outside Earth's effective gravitational field. The continents are mounted on plates that float on a sort of lava. Arrhenius's charged ions are kept apart by insulating coats of water molecules.

Resistance based on impressive yet erroneous arguments is no thing of the past. When geology was sure that all change had occurred gradually over long periods of time, that was excellent reason to label "preposterous" and "incompetent" J. H. Bretz, who explained various aspects of the Pacific Northwest as resulting from colossal floods during the last Ice Ages; but fifty years later Bretz was honored by the Geological Association for those very same insights (Allen et al. 1986). Not so long ago Carl Woese showed by laborious analysis of RNA that the tree of life is rooted in the Archaea (Morell 1997); but the first reaction to his revolutionary demonstration was that "his tiny snippets of tRNAs were . . . too fragmentary to be reliable indicators of evolutionary relationships" (701–2)—an eminently reasonable view, which, however, turned out to be irrelevant in this instance.

Inability to explain is never a conclusive reason for asserting that a phenomenon does not exist. So the best answer anomalistics can give disbelievers is to point out that explanations have always come only in their own good time, even for quite undeniable natural phenomena. But instead of sticking with observations and empirical fact, many anomalists counter such demands of "How could . . . ?" by attempting to suggest possible mechanisms for the claimed effects. That is generally self-defeating. So long as one is only claiming that an anomalous phenomenon occurs, one is making a crypto claim, a simple empirical existence claim, which is not in itself a radical contradiction of established knowledge even if debunkers assert it to be. The history of science brims with discoveries of entirely unexpected facts, and recent years have seen more rather than fewer of those—say, the realization that carbon can exist, and does in nature, not only as diamond and graphite but also as bucky-balls. However, attempts to explain how something might happen that is contrary to contemporary beliefs is a *para* claim, the assertion—or at least suggestion—of a theory that has no other basis than the disputed facts themselves. So implausibility is heaped on improbability, which is no way to convince an unbeliever let alone a disbeliever. For example, Benveniste has suggested that the active allergen molecules somehow imprint their activity in some electromagnetic fashion on the water molecules—relying on several effects involving electromagnetism and chemical molecules that are unknown to physics or chemistry. It's altogether a mistake to attempt to explain an anomalous phenomenon before its very reality has been accepted.

That is to say, it's a mistake in terms of the argument, in terms of presenting the anomalist case to nonbelievers and disbelievers, where the concern is, how best to go about establishing credibility in the face of incredulity. In intellectual terms and within the anomalist community, it is by no means a mistake to speculate about possible mechanisms; indeed it may be a useful guide to further research. In mainstream science and scholarship, too, the overriding value of theories lies in guiding research by suggesting possibilities, so that even farfetched notions are allowed, as in cosmology the "many-universe" "explanation" of why *our* universe has just the right parameters for us to be here.

Arena of Dispute

Whereas most issues in science (issues of knowledge by contrast to application or policy) are argued just among the active researchers, anomalistics is typically argued in the public domain. Most of what is written and said about New Age scientific heresies is directly addressed to the general public.

Psychic phenomena, telepathy, ghosts, and the like are part of the popular culture, learned through children's stories, folklore, and myth. So the general public is comfortable hearing and talking about these things without the benefit of expert interpreters. Science, by contrast, is learned (at best) only at school, and few people reach the stage of feeling that they can understand anything about science without the help of specialist interpreters.

The debunkers who want to call on "science" to discredit unorthodoxies like parapsychology thus find themselves trying to make the general public believe that science requires them to jettison familiar beliefs just because science supposedly tells them to. Now people have learned to accept from science notions that contradict common sense on matters of science: that time slows down for an electron when it goes fast enough, say; but people are less likely to accept notions that contradict common sense about everyday matters, which include premonitions, intuitions, and coincidences, whose personal significance seems self-evident and which we *know* happened to us.

Scientists and scholars tend to have little experience of public debate, and they rarely perform well at it. They are accustomed to the rules of internal disciplinary argumentation, where substantive issues are the clear focus of attention and the participants concern themselves only secondarily with how best to make their points. Scholars and scientists have relatively little occasion to develop expertise in countering unscrupulous innuendo, deliberate obfuscation, or uninhibited ad hominem attacks. They may then do miserably in public arguments, recognizing only after the event that the debate was primarily ideological and only symbolically about the nominal topic. Those who debate creationists have had this experience more than once, notoriously in the televised encounter in 1981 between the creationist Duane Gish and the biochemist Russell Doolittle (Lewin 1981).

Tactics

Ethics

In some areas of social science, the ethics of misleading the subjects under study is argued perennially because some research cannot be carried out without a measure of deception (Coughlin 1988).

Misleading those who do the investigating is another matter. In mainstream research one is expected not to attempt to win technical arguments by hoodwinking one's opponents; in anomalistics, however, deception is seen by some as a necessary and appropriate way of making a point. Early in this century, a physicist used deception to demonstrate that N rays were "pathological science," not genuine science. In the 1980s, Randi's Project Alpha (Randi 1983a,b; Truzzi 1987) used a couple of young sleight-of-hand artists to pose as psychics to gain entry into a parapsychological laboratory and mislead the investigators. Thus ends are held to justify means. But fooling one laboratory hardly discredits a whole field. Incisive critiques of debunkers' tactics have been rendered in the Rockwells' "Irrational Rationalists: A Critique of *The Humanist's* Crusade against Parapsychology" (Kurtz 1978; Rockwell et al. 1978a,b) and in Hansen's "CSICOP and the Skeptics: An Overview" (Hansen 1992).

Making the Substantive Case

It takes much longer to explain why a point is erroneous than it took to assert the point. It can be very tiresome to answer in full detail what seems like a poorly based, incoherent case for something highly improbable. Few scientists think it worthwhile to spend so much time on a task that shouldn't need doing, if only the public were more scientifically literate. The frustrations of arguing with a crank have been described with feeling by some who have had or witnessed the experience (Russell 1956; Shaw 1944:360–61).

Drawn into dispute, frustrated experts may become arrogantly dismissive, giving short shrift to what seems like incompetence or nonsense. But that leaves them open to the charge of dogmatism and arrogance and they lose debating points and public credibility.

Such displays of arrogance and dogmatism often become the basis for a broader critique of allegedly excessive conservatism on the part of Establishment science (Bauer 1984a:57–61). That in turn is likely to reinforce the experts' view that it would be a waste of time to attempt a scientific, substantive debate. It places the experts in the position of having to defend science in general. In that role they will not find it easy to say, "I don't really know for a fact; but based on what we do know in this field, the best guess would be that . . . Bigfoot are very unlikely to be prehistoric survivors of *Homo erectus,* Uri Geller cannot bend keys just through mind-power, UFOs are not extraterrestrial visitors." Rather, the polarization of argument and lack of substantive common ground provokes statements like "There's no such thing as 'mind-power,'" or "UFOs are definitely not alien visitors."

Thus the experts *do* speak in dogmatic tone. Since everyone knows that science ought to be anything but arrogant or dogmatic, true believers in anomalies gain public support by thus showing Establishment science to be breaking its own rules. Bystanders who may not feel comfortable with the technical issues are quite capable of discerning arrogant dogmatism and wishing those who practice it to be on the losing side.

Appeal to Authority

Appeals to an ideal "science," which are made by both sides, are analogous to the appeals often made in public disputes over sociopolitical matters to such universal values as "justice," "democracy," or "patriotism": both sides agree on the principle, but each side has a different interpretation of how that general principle applies to the specific issue at hand.

There are, after all, few if any generalizations that don't call for exceptions to be made in practice. Even "motherhood" isn't so wonderful if the mother is an unmarried twelve-year-old rape victim. Patriotism was not necessarily admirable in Nazi Germany. "Open-mindedness" in science is good only under some circumstances and not under others: one ought not to waste time being open-minded about the Earth's maybe being flat, say; as G. K. Chesterton pointed out, the purpose of an open mind is the same as that of an open mouth—that it may close again on something solid.

A common but noteworthy version of the appeal to science by debunkers runs like this: "Science should always be believed, because science is self-correcting and changes its mind as called for by the evidence."

Unwary bystanders might swallow this rhetorical trick instead of realizing that the issue then becomes, is this particular anomaly perhaps one of those over which science will change its mind?

Galileo and Fulton

A characteristic counterattack against the mainstream is for anomalists to assert affinity with people who were once derided by the Establishment but who were later vindicated. Writing in 1842 about homeopathy, Oliver Wendell Holmes remarked that advocates of a "delusion" always compared their persecution to that of Christ, Copernicus, and Galileo and branded their opponents as like the Holy Inquisition.

Anomalists also like to invoke Kuhn's notion of paradigm shifts: before the shift, the mainstream simply doesn't comprehend the unorthodox claim; and they can cite a number of anomalies that were eventually vindicated. But arguing for legitimacy in this manner is also a misguided appeal to authority, described once as "the Fulton non sequitur" (Gruenberger 1964): "They laughed at Fulton, but he was right. They laugh at me, *therefore* I am right." For every scientific unorthodoxy later vindicated there are a host of others, not now remembered except by a few aficionados, where the unorthodoxy turned out to be a false trail. There is no ready source of illustrations of all the errors that science has eliminated over the last few centuries, and so we're not sharply aware of the fact that very few of the anomalous facts or unorthodox ideas that crop up all the time in science ever turn out to be valid.

Oversimplification

Oversimplification is commonplace. Issues become black-or-white. Though in recent times there have been some instances in parapsychology of skeptics and confirmed believers collaborating (Bem and Honorton 1994; Wiseman and Morris 1995) or suggesting collaboration (Polidoro 1997), still most of the literature in anomalistics is really two side-by-side genres, the pro and the con.

Rarely if ever is anomalistics given credit for grains of truth. Velikovsky was and is said to be "wrong"—a global judgment when in reality he made a host of assertions, some of which had a measure of validity. One arguably less-than-competent laboratory is asserted to typify all of parapsychology, whereas one less-than-competent forensics laboratory is hardly taken to show that forensics is pseudoscience. Reich is called a pseudoscientist because of his work on cloud-busting and bions, yet his views on the relation between psychology and politics are far from obviously invalid and his psychological-somatic therapy continues to help some number of clients (Edwards 1967, 1974).

Questions and Excuses

Rhetorical questions abound. Against parapsychology, "Why has there not eventuated a reproducible observation or experiment after so many years of trying?" "If that group of psychics was able to make some money on the silver market, why has it not continued to amass wealth, if only to provide the funds for research that parapsychology clamors for?" "How could bones of Bigfoot not have been found if they exist, with so many people finding footprints of them?" And so on and so forth.

Once a given issue is settled one way or the other, answers to such questions will be evident enough; indeed they are likely to appear obvious in hindsight. Before the issues are settled, however, the inability to provide conclusive answers proves nothing.

Somewhat akin to this rhetorical questioning, in that it is pure speculation, are some standard excuses; in the case of Nessies, for eyewitnesses' photos that didn't pan out: cameras jammed, the lens cap had not been removed, the battery was dead, the film was all used up. True believers may accept these explanations, but the disbelievers find such excuses ample ground for further discounting the eyewitness reports.

Unwelcome Bedfellows

In all controversies there is polarization and a tendency to force participants into one of two camps. People then find themselves willy-nilly in the company of others with whom they really have nothing much in common. Excluded from mainstream discourse, anomalists find themselves without obvious means to control who is in their ranks. Whoever alleges to be a ufologist is thereupon taken by the media and the public actually to be one, even if most ufologists regard the aspirant as unqualified, insincere, fraudulent, or generally undesirable. Serious anomalists and hardheaded semibelievers find themselves associated with all sorts of alternative-energy and alternative-science buffs; with academics and scientists who lack the competence to make a mark in the mainstream; and with inveterate anti-Establishment types and the perpetually angry and paranoid. There are overlaps between anomalistics—heresies as to knowledge—and minority groups in such other spheres as politics and sexual behavior (I discuss this in chapter 7).

In science, and in academe generally, the norm is to trust colleagues, the literature, and even competitors to be honest and reasonably competent, unless or until there are strong grounds for distrust. Scientists who become involved in anomalistics are naively unprepared for the range of personalities and unreliability one encounters there. A notorious example is the succession of sleight-of-hand "mediums" who have, for more than a century, periodically hoodwinked scientists seeking to study psychic effects. During the surge of interest in cold fusion, a prominent chemist found himself gulled by a professional charlatan (Begley and Levin 1994; Pool 1993b).

I was once on a panel of academics inquiring into allegations that a chemistry graduate student had faked data in his Ph.D. thesis. The panel's chairman was a professor of law who had been a practicing lawyer and had also served as a judge. The student was interrogated for several hours, and it was evident from the questions around the table that opinion among the academics was quite uncertain. The chairman, however, as early as the first lunch break, was unequivocal in his view that the student was lying—as all agreed several days later. As a lawyer and a judge, through much contact with liars, that chairman had learned much about assessing credibility; we other academics had lived in an environment much more sheltered from liars and cheats and were not familiar with telltale clues that experienced lawyers and judges readily detect.

Guilt by Association

Inferring guilt by association is often labeled improper and invalid. When totalitarian regimes hold children responsible for the sins of their parents, or vice versa, it's easy to see how wrong guilt by association is. The father of a criminal may not himself be criminal. That type of inference is wrong because it associates different kinds of things, familial relationship and obedience to the law. On intellectual questions, however, it is surely legitimate to consider intellectual associations as meaningful; indeed, the coherence of modern natural science provides great justification for insisting that new science should fit seamlessly with older stuff.

Debunkers quite typically seek to assign guilt by association. It's no easy thing to discredit entirely the major anomalist claims by careful discussion of the evidence. It's much easier—and so it's done all the time—simply to include them all in the same list, as equally "pseudoscience." But this lumping also has disadvantages. It weakens the intellectual case: newspaper horoscopes are obviously a very different matter than the experimental protocols described in peer-reviewed articles in parapsychology. Furthermore, rejecting as wrong something that is merely unproven, and then linking disparate topics of that sort together, can backfire if even one of the unorthodox claims turns out to be valid, as some do. For decades there were those who decried as wasteful, or worse, the use of vitamin supplements, but they will (or should) have been mightily abashed when in 1998 the Institute of Medicine recommended such supplements even for people enjoying an apparently adequate diet.

Debunking loses credibility when it calls "paranormal" or "supernatural" the search for such entirely material albeit as-yet-uncaptured species as the giant sloth; the searches may be for something that never existed or isn't extant and therefore perhaps misguided, but surely not paranormal. Debunkers often cite their concern for public rationality and scientific literacy; but by their lack of discrimination, and by their doom-saying and exaggerated assertions of the harm that supposedly flows from what they call pseudoscience, they fail to practice the rationality and scientific approach they preach.

Guilty; But of What?

The critics of pseudoscience are inconsistent in the criteria they rely on to label subjects "pseudo." Things like polywater are held to deserve being called "pseudoscience" because they're known to be wrong. Yet the same pundits will insist that science is not characterized by always being right, or in any other particular result, but only in the process of using the scientific method. Yet when right results are obtained by people who flout the scientific method and other norms of science—as with high-temperature superconductors—their lapses are not criticized. It's like the man who was known to be guilty while it remained to be decided, of *what:* anomalistic pursuits can be criticized for following improper procedures, or for getting to a wrong place, or for studying something that's not worth studying, just according to the specific case that the science-police happen to have on their minds (or in their craws) at the time.

Evidence

Anomalistic disputes have to do with evidence in characteristic ways.

One is the already mentioned fact that anomalistic mysteries don't fall neatly within the purview of any of the established intellectual disciplines. In fields that are interdisciplinary or multidisciplinary—or in anomalist cases often *un*disciplinary—there are no universally agreed criteria for what constitutes good evidence. Good practice in history may not be the same as good practice in geology or in folklore studies; what one has to do in seeking living dinosaurs in the Congo may not seem like good practice to people in other fields of endeavor.

In the anomalistics literature, all too often insufficient documentation is given to allow others to check the sources. Even tangible evidence is problematic in anomalistics as it usually isn't in science. There exist no reliable, accredited repositories or museums of ufology or cryptozoology, so specimens or artifacts mentioned in the literature often cannot be retrieved for reexamination. Over the centuries there have been a respectable number of reports of stranded carcasses of sea serpents or sea monsters belonging apparently to no known species (Heuvelmans 1968); but in most cases, only sketches or photographs or eyewitness descriptions of them are now available. Even when tangible specimens *are* produced, their provenance is questionable. Material that has been passed from one private citizen to another doesn't carry the same assurance of authenticity as something from the vault of a museum, especially on matters in which fraud or hoax is a constantly present possibility. A video of the autopsy of an alien was received in the mail by a ufologist from an anonymous source; how much credibility can or should that be granted?

In an established field, it's usually easy to decide whether something is relevant or not. For chemical reactions, one need not take into account the position of Venus in the sky nor the atmospheric pressure outside the isolated experimental system. It's known with a very high degree of certainty that those things don't have to be taken into account in chemistry (the claims of Piccardi [1962] notwithstanding), because so many concordant studies have achieved the same results without controlling those factors. But in telepathy or clairvoyance, who is to say what factors might be relevant?

Wherever a claim is being supported by statistical argument, it is relevant to the estimation of statistical significance how often the experiment has been tried *unsuccessfully,* anywhere at all. There is the "file-drawer effect": if only successful experiments are published while unsuccessful ones remain in files in the investigators' desks, then it may be thought that, say, the psychic effect is observed seven out of ten times when the actual ratio might be seven out of a hundred, or a thousand, or an even larger number—which would make it more likely that the successful trials were flukes.

The Evidence Cuts Both Ways

In anomalistics, where by definition the evidence is not utterly compelling, believers and debunkers are thereby free perpetually to reach opposing conclusions, to fit the evidence into their opposing stories:

- To anomalists, the unexplained residuum is the real evidence. To the deniers, this is like looking at the small residuum of cases where a grown-up is *not* caught putting the presents from Santa into the Christmas stocking or the tooth-fairy money under the pillow.
- In medicine, the unexplained residuum tends to be ascribed by the "alternative" school to faith healing, therapeutic touch, or the like, whereas to the conventionalists it is owing to the placebo effect or spontaneous remission.
- Concerning yeti or Bigfoot, and the fact that apemen are featured in folklore across the world, Bayanov has pointed out that reported sightings are either mistakes influenced by expectation based on the folklore or that such creatures exist in folklore because human beings have encountered them over the ages. In fact, "the existence of mythological hominoids is a necessary, though not sufficient, condition of the existence of real hominoids" (Bayanov 1982). Their absence from folklore would even speak against the creatures' existence, which is the opposite of the debunkers' usual argument.
- In the case of memories of alien abductions recovered through hypnosis (Jacobs 1992), while the nay-sayers suggest that the similarity of the accounts is explained by unconscious suggestion on the part of the hypnotists, of course the accounts would also be similar if the abductees were simply reporting the truth about a shared experience.
- Concerning Loch Ness monsters, there's no dispute over certain facts; yet each of those "facts" can be quite plausibly used in support of the diametrically opposite views that Nessies do not exist and that they do exist (chapters 1 and 2, respectively, in Bauer 1986).

Probability and Statistics

Assessments of plausibility are central issues in anomalist controversies. Many arguments are expressed in terms of the odds against something's happening by chance. But those estimates are frequently unwarranted. Most people not schooled in statistics or in probability theory have a very bad intuition about the likelihood of certain sorts of events, for example "coincidences."

Coincidences

When I met at Loch Ness a fellow Nessie fan with whom I had corresponded for some time, it turned out that his mother's first husband had been a statistician who later emigrated to the United States—and whom I had come to know reasonably well just a few months earlier as he was a visiting scholar at my university. What a coincidence!

The same Nessie fan introduced me to Winifred ("Freddie") Cary, who lived close by and had seen the monster many times. One of her neighbors turned out to be an American writer who had been educated at Hollins College (known since 1998 as Hollins University), in Roanoke, the city closest to my university. What another amazing coincidence!

And then there was the student physiotherapist from Aberdeen, Scotland, who attended my best friend in Inverness and turned out to have been in one of my chemistry classes ten years earlier in Virginia, USA. Another stunning coincidence.

But are these coincidences really so amazing?

At my university, I had come to know dozens, scores, indeed hundreds of people and over the years had taught thousands of students. I had also known and taught further hundreds at other places I had lived and worked. At Loch Ness on any of my visits there, I encounter a few dozen people with whom I talk more than casually. There are many possible sorts of relationships to provide "stunning" coincidences—beside mother's first husband, *any* sort of relatives; some sort of business relations; acquaintances through hobbies or schooling; location of someone's place of education or business or vacation; and more. So there were an enormous number of potential coincidences: thousands of people I knew from various places, multiplied by a few dozen talked to at Loch Ness, multiplied by hundreds of possible relationships or places. There are millions of possible "chance" concatenations, so among all those possibilities some sort of coincidence is quite likely to occur whenever I am at Loch Ness. Even more likely is it that some sort of coincidence happens to me at some time in some other place (Good 1966, 1980).

When a coincidence comes about, we are struck by how remarkable it is; yet it would be much more remarkable if we never experienced any coincidences at all. However, we tend to take as personally significant anything that happens to us; we often fail to realize that it could so easily occur for no good reason other than chance. The odds against my winning the lottery are very great, though I have a ticket in each drawing; but *someone* wins the lottery almost every time. And some of those winners no doubt imagine that their win is somehow personal, typically because they had a hunch that this time they should buy a ticket, or because they chose the right numbers. Far from it: the numbers chose *them*.

Many anomalist claims are based on misunderstanding of the likelihood of coincidences, for example finding "keys" in Shakespeare's plays that reveal who the "true" author is (Friedman and Friedman 1957) or noting similarities between the assassinations of Abraham Lincoln and John Kennedy (Reader's Digest Association 1982:65–66).

Calculating the Odds

There's the apocryphal story of the man who received unsolicited in the mail from a stock-market tipster a "free sample" predicting that next week, stock A would go up. It did. The following week he received another such tip, this time predicting it would go down; and again the tip proved correct. Week after week the tipster was right. After ten weeks, no free sample was enclosed but rather an invitation to subscribe to the service, at one hundred dollars per year; quite a bargain, given that the service had proved its accuracy during the ten-week period.

But no one can predict stock-price fluctuations with that accurate consistency.

Yet anyone can generate such a string of successful predictions.

In the first week, send out 1,024 "free samples," 512 predicting "up" and 512 "down." In the second week, to whichever of the two groups of 512 turned out to have been sent the correct prediction, send 256 "up" and 256 "down." And so on. After 11 weeks, one person will have had 11 straight weeks of correct predictions. That person will most likely have become quite convinced that the tipster understands the market superbly well.

Thus taking things as personally significant together with untrained intuition about probabilities leads us astray. Knowing that a penny has equal chances of heads and tails, one might expect a fairly steady alternation of heads and tails when a coin is tossed: in other words that the succession HTHTHTHTHT is more likely to happen than HHHHHHHHHH or TTTTTTTTTT. Naive expectations of this sort allow casinos and professional gamblers to make a very good living, for they understand about odds as most of the rest of us do not. In fact it's just as likely that a coin will fall HHHHHHHHHH as that it will fall HTHTHTHTHT. To get the latter, there is 1 chance in 2 that the first toss is H; the same that the second is T; the same that the third is H; and so on. The probability of the sequence is ½ multiplied by itself ten times. For HHHHHHHHHH, exactly the same calculation applies.

People who take HTHTHTHTHT to be a normal expectation through chance are confusing *relative frequency* of heads and tails with the probability of a particular *pattern* of heads and tails: of course the chance is considerably greater that in a total of 10 throws, there will be 5 H and 5 T rather than 10 H or 10 T, because the equal division can be obtained in a number of different ways: HHHHHTTTTT; HHHHTTTTTH; and so on.

It's widely believed that athletes experience occasions when everything goes right for them, as when basketball players have a "hot streak." However, keeping track of throws made and missed by a given player reveals that these "hot streaks" occur with the same frequency as they would by chance; chance demands that *some* player, a certain proportion of the time, will score with an "unusual" number of throws in a row—and also at other times that player will miss with an "unusual" number of throws in a row.

A commonly used demonstration of how wrong our intuitions are about chance is the Birthday Problem: how many people must there be present to give a 50:50 chance that two have the same birthday?

Since there are 365 days in the year, many people guess about 183. But no matter how they answer, most people are surprised that the correct answer is as small as 23. Yet that's easy to understand once you know how to think about it. The chance of two same birthdays depends on the number of *pairs* of people present (Good 1978:343); and the number of pairs of people increases much more rapidly than the number of people present, as shown in table 3.

Table 3. Pairs Increase Rapidly with Number of People Present

Number of people	1	2	3	4	5	6	10	15	23	25
Number of pairs	0	1	3	6	10	15	45	105	253	300

Of course it's quite different if you ask, How many people must be present to give a 50:50 chance that someone has the same birthday as you (or any other specific person or date)? That is 253, much larger—because all the pairs are then formed between yourself and other people, and the number of pairs then does increase in proportion to the number of people. The difference between the earlier 23 and the present 253 is very roughly analogous to the interpretation of coincidences: some coincidences happen all the time, but any given coincidence will be relatively unlikely.

When underwater photos seemed to show a Nessie, Robert Rines and Sir Peter Scott assigned the creatures a scientific name, *Nessiteras rhombopteryx:* Greek for "the monster of Ness with a rhomboid-shaped fin" (Anon. 1975). Almost immediately it was pointed out (Meecham 1976) that this name is an anagram for "Monster hoax by Sir Peter S." Obviously Scott must have chosen the name with this hidden clue in mind, for the chances against such a coincidence seemed astronomical. A journalist who "asked a Leeds mathematician and statistician the odds on such a coincidence . . . was told, 'Technically, they may be untold-millions to one. About the same as the chances of the Red Sea parting tomorrow.'" An expert with odds turned to a more practical use, a spokesman for Ladbrokes, the betting establishment, thought the anagram was "totally unbelievable. I really don't know what to say. If this is a coincidence, the odds on it occurring by chance are in excess of a million-to-one" (Coggins 1976:58).

But those experts were quite wrong. The letters of the alphabet don't occur with equal probability in a given word, phrase, or sentence, as addicts of word games like Scrabble or Perquackey know. Any combination of letters that has made one phrase is thereby likely to contain the more commonly used letters, and so is quite likely to yield other phrases as well. It turns out that *Nessiteras rhombopteryx* has quite a lot of anagrams of possible relevance (Meredith 1976). Rines and Scott themselves responded with, "Yes, both pix are monsters. R."

Many then expressed surprise at the number of different, apparently meaningful combinations one can find among such a small set of letters. When the number of available letters and combinations of letters becomes large, say, in a book like the Bible, the chance for "meaningful" surprises becomes very great.

The abstract argument that such hidden messages can arise by chance is hardly convincing, however, to most people who aren't versed in combinatorial mathematics. The most telling demonstration then is not to lay out the mathematics but to demonstrate how easy it is to recover messages by similar decoding in other texts: for example, the King James version of Genesis (about 150,000 letters) delivers several combinations of "Roswell" and "UFO"; as well as "comet," "Hale," and "Bopp," together with "forty" and "died"—surely the clearest possible message about the Heaven's Gate mass suicide (Thomas 1997).

When urged to believe something because the odds on it are, say, a million to one, it's good to be aware that even expert statisticians argue over how such odds should be calculated. There are at least two major schools of thought, the non-Bayesian and the Bayesian, and the difference can be striking.

Non-Bayesian Approach

The non-Bayesian approach is much more widely used than the Bayesian, especially but not only by those not expert in statistics; but it underestimates the probability of improbable events.

The non-Bayesian method compares results with what would be expected to happen "by chance" and then calculates the probability of any discrepancy. The trouble is, that calculation depends on *the distribution:* how frequently actual results will fall how far away from the "true" value. But often one doesn't know what the distribution is—one is dealing with phenomena not well understood, after all. So one has to postulate a distribution.

Among distributions, many nonstatisticians have heard only of the "normal distribution" or "bell curve." But there are many situations in which that is inappropriate: for instance, in the once-popular card-guessing studies of extrasensory perception. A pack of 25 cards ("Zener" cards) contains 5 cards of each of 5 symbols (cross, box, star, circle, wavy lines). By chance, one guesses on average 5 correctly in each run of 25 cards: 1 chance in 5 each guess, for a total of 25 guesses. To interpret nonaverage scores—12 out of 25, 3 out of 25, whatever—one needs to know how likely each of those scores is "by chance." Clearly that's not a symmetrical bell-curve distribution, because there are only 5 possible scores below average (0, 1, 2, 3, 4) but 20 possible scores above 5; the chance distribution is skewed (see fig. 1). Even a quite good book (Shermer 1997:71) devoted specifically to explicating how to think about anomalies gets this wrong and shows a symmetrical bell curve for this situation.

If one uses a bell curve instead of the correct distribution, one will calculate wrong odds that a certain thing might have happened by chance. There are innumerable situations for which the proper distribution is not a bell curve but rather an unsymmetrical one, sometimes very heavily skewed; statistics

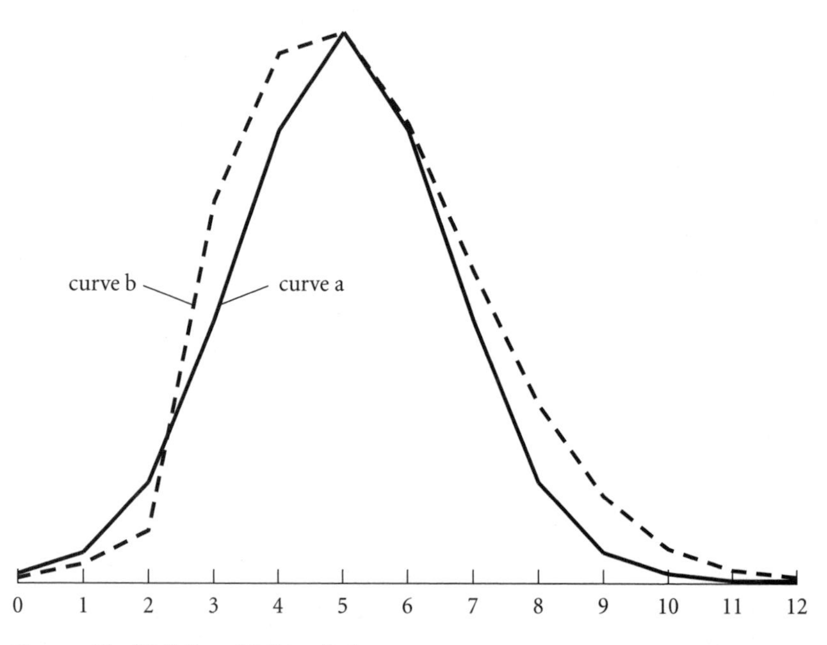

Figure 1. The "Bell Curve" Misapplied

The well-known bell curve *(a)* is symmetrical, but the probabilities of many events in the world are *not* symmetrically distributed; for instance, the card-guessing trials described in the text are distributed like curve *b*. If a bell-curve distribution is erroneously assumed for such a case, then the conclusions drawn can be wildly misleading.

knows of a number of distributions encountered in the world other than the "normal" one.

Moreover, though it might seem intuitively obvious what "by chance" means, there are deep and unresolved issues here. It may seem natural to equate "by chance" with "random," yet there is an important difference (Good 1953, 1974). As Seife (1997) points out, "In scientific experiments such as drug trials, researchers randomize the test subjects to avoid bias. But randomizing by coin toss or luck of the draw can occasionally produce an orderly pattern—with all the women assigned to the control group and all the men to the study group, to take an extreme example."

In other words, doing something "at random" like tossing an unbiased coin or choosing blindly from a pile of objects does not necessarily in any given trial produce a "random" or "by chance" result: a bridge player occasionally is dealt all the cards in a given suit; coin-tossing occasionally yields a run of ten heads or tails. A mathematical test for degrees of randomness turned up some intriguing illustrations of this. Intuitively one might expect that the sequence of numbers would always be random behind the decimal point of an irrational number—a number that has no exact decimal equivalent so that decimal figures continue infinitely after the decimal point; yet the randomness turns out to be greater with $\sqrt{2}$ than it is with *e*, the base of natural logarithms, which in turn shows more randomness than does $\sqrt{3}$ (Seife 1997). This is of more than purely academic or mathematical significance: experimenters, including in parapsychology, have sometimes used the decimal places of irrational numbers as a source of supposedly random sequences of numbers.

Bayesianism

For an introduction to the Bayesian approach see, for instance, the article by Jefferys and Berger (1992), which uses as an example the commonly invoked criterion for being scientific, namely Ockham's Razor—choosing the simplest available explanation. That article also contains references to standard works on Bayesian and non-Bayesian statistics.

Very roughly speaking, Bayesians begin with an estimate of prior probability, which may be simply an informed guess: say, that the probability of a coin falling "heads" is 0.5, or that the probability of someone being able to see into the future is a billion to one against. As relevant experience accumulates, that prior probability is progressively modified, using the results of observations and experiments to estimate the weight of evidence—the probability that the result would have followed if the hypothesis (the estimate of prior probability) had been correct. In the case of the coin, a very long series of results very far from 50 percent heads would be required to change the estimated probability from 0.5. For looking into the future, even the startling success of a few predictions would change the odds only from one in a billion to perhaps one in hundreds of millions (Dobyns 1992; Dobyns and Jahn 1995; Jefferys 1990, 1992, 1995). The Bayesian approach, in other words, satisfies the common aphorism that extraordinary claims require extraordinary proof; in other respects too it seems well suited to the issues arising in anomalistics (Good 1992; Sturrock 1994).

Another paraphrase of the import of Bayesian interpretation puts it thus: the actual occurrence of a highly unlikely event does not necessarily make that event less unlikely. The experience of any given coincidence doesn't make that particular coincidence any more probable. That some bridge players have actually been dealt hands containing all thirteen cards of a given suit doesn't change the astronomical odds against its happening. That I may have won the lottery leaves the chance of doing so exactly the same as before I won it.

The difference between inferences drawn from Bayesian and from non-Bayesian bases can be striking; and it is especially germane to the question of establishing the reality of an unlikely (anomalous) claim. The standard non-Bayesian approach is "capable of suggesting significance in results [that are] in fact due to nothing more than chance" (Matthews 1999:1). Table 4 shows how different is the required *p*-value under the Bayesian and non-Bayesian approaches if one wants to get odds of 20:1 that a claimed phenomenon exists.

The first two columns describe the level of initial expectation, the prior probability, which is a choice made by the investigator. (That this is subjective is a criticism often leveled at the Bayesian approach. One obvious response is that the non-Bayesian approach is also subjective, most obviously in making the choice of a particular distribution.) The third column is the two-tailed *p*-value corresponding to the requirement that the Bayesian calculation yield odds of 20:1 that the experiment demonstrates the claim to be valid. The last column is the ratio of that to the commonly used non-Bayes interpretation that a *p*-value of 0.05 represents odds of 20:1 that the claim is valid.

Table 4. Bayesian vs. Non-Bayesian Estimates of Odds

Level of Initial Skepticism (*Qualitative* Expression)	Pr (*Quantitative* Expression of Level of Skepticism)	Maximum *p*-Value to Indicate That Anomalous Claim Is Probably Valid at Odds of 20 to 1	Overestimate of Odds by Non-Bayes Approach
Agnostic	0.500	3.0×10^{-3}	17
Mildly skeptical	0.900	2.0×10^{-4}	250
Moderately skeptical	0.990	1.3×10^{-5}	4,000
Highly skeptical	0.999	1.0×10^{-6}	50,000

In other words, if you believe there is a 50:50 chance that telepathy might happen, and you design a series of experiments to test that belief, under the Bayesian approach the results would have to indicate that only 3 times in a thousand could they be obtained by chance, in order to make the odds 20:1 that telepathy indeed was at work. The non-Bayesian, by contrast, interprets the "3 in a thousand" as odds of 1000:3 that telepathy is real, an overestimation by a factor of about 17.

The significance of this for people investigating anomalies, which is to say unusual events, is that the overestimation becomes more serious, the more improbable the phenomenon is judged to be. The importance for all of us lies in the possibility that many loudly touted claims for the effectiveness of new drugs or surgical treatments are badly overstated because of reliance on incorrect statistical interpretations (Matthews 1998).

Popular Misinterpretations and Cultural Probabilities

Intuitions about probability tend to be bad also through ignorance of certain regularities of human culture. When a performing magician asks a volunteer from the audience to think of a number between 1 and 10, the audience presumes that there's an equal chance of thinking of a "1," a "2," a "3," and so on; but the magician knows from experience that this is not the case: about two-thirds will choose "7" while very few choose 1, 2, 4, or 10—presumably because in our culture "7" and "3" are somehow "special" whereas 1, 2, 4, and 10 are not. Knowing such "population stereotypes" enables some people to hoodwink the rest of us who don't know them (Marks and Kammann 1980).

In considering what "chance" could bring, we also characteristically neglect to consider all possible outcomes. For example, we've picked a card from a pack, it turns out to be the eight of hearts, and the magician on stage amazingly enough pulls out of his breast pocket a large piece of cloth imprinted like an eight of hearts. How could he have foreseen this? Surely this is psychic precognition?

But that stage magician may have in his many pockets many different items that, if we had drawn some other card out of the pack, could have been dramatically revealed with equally surprising effect. Moreover, we were not told beforehand that he *would* pull such an item out of his pocket—had we drawn a card for which he had no suitably surprising item, he would simply have omitted this "surprise" from this particular performance. There was nothing surprising about that cloth eight of hearts; it shows only that the magician was well prepared to perform a variety of illusions.

The non-Bayes calculation of "odds against chance" suffers a similar defect: the presumption is that a particular result was obtained either by chance or because of the claimed anomalous effect, such as *psi,* the postulated cause of psychic effects. But that assumes that the experiments have satisfactorily controlled for everything else, which the state of human ignorance renders unlikely, particularly with slippery phenomena.

Not only true believers but debunkers too may be wrong about probabilities, and may label as "superstition" real things that seem intuitively wrong. The fact that you "never" seem to join the line that moves fastest, for example, is *not* a popular superstition but quite realistic: if there are only two lines, you have a 50 percent chance of choosing the faster one, and your chances only go down as the number of lines, of possible choices, increases. There is an entirely scientific basis for Murphy's Law (Matthews 1997).

Everyone should read the classic exposition, written for the general reader, *How to Lie with Statistics* (Huff 1954). For those willing to follow some mathematics, *Lady Luck* (Weaver 1963) may be a congenial introduction. Fallacies "are not the prerogative of fools and statisticians," as I. J. Good (1978:337) points out in discussing a host of errors relating to factors such as sampling or analysis.

Typical Fallacies

Wherever people argue, there are likely to be fallacies. I. J. Good (1962a) lists several dozen types under such headings as

- Psychological fallacies: suggesting through implication of prestige, appeal to authority, and the like; emotional reasoning, for example the use of emotionally toned words (we are steadfast, the opponents are stubborn);
- Linguistic fallacies: making a statement that implies "all" when "some" is true, as in stereotyping—"red-headed people have bad tempers";
- Change of proposition: confusing the issue, diverting to another question, bringing in red herrings;
- Oversimplification; including "cliché thinking" like "everything is relative" as a summary of the theory of relativity;
- Logical fallacies: non sequitur, arguing in a circle, begging the question, smuggling assumptions into an otherwise logically correct argument.

Just knowing about these categories is not in itself enough to prevent us from committing the fallacies; anomalist controversies—like all arguments—are replete with mistakes committed by people who know full well that those types of errors exist.

It's easy, for example, to understand why it is fallacious to argue backward from an effect to identify what went before as its cause (argument a posteriori, or after the fact). Yet the error is committed over the origin of life by such eminent people as Sir Karl Popper, Sir John Eccles, and Sir Fred Hoyle. The probability of all the atoms lining up "by chance" to produce the DNA in anybody's cell is calculated as perhaps 1 in $10^{100,000}$ or so; and the conclusion is jumped to that it couldn't have happened "by chance," some intelligence must have designed it that way. But DNA did not suddenly appear out of a random collection of atoms; still less did any particular person's DNA so appear. Innumerable *nonchance* chemical and other interactions together with *some* chance events, over several billion years, brought into being our present stock of DNAs. "The random possibility of *anything* happening in the particular way it *did* is pretty near impossible beforehand, but once it *does* happen, the odds in favor of it *having happened* are 100 per cent" (Estling 1983:619).

Good (1962a) lists dozens more fallacies pertaining specifically to probability and statistics, and includes a useful bibliography of works on logic and statistics. Here I shall focus on fallacies that are commonly found in arguments in anomalistics.

Believers

Some fallacies regularly emanate from proponents of unorthodoxies:

- That an accumulation of data that is in some manner flawed—lack of controls, not reproducible, in the right direction but not statistically significant—adds up to solid support if only there is enough of it.

  One finds people saying that so much evidence has accumulated over the centuries that extrasensory perception occurs, that it becomes foolish to deny it. Yet asked for strong, exemplary evidence, parapsychologists disagree among themselves as to which evidence is the best. John Beloff, one of the grand old men of parapsychology, listed "Seven Evidential Experiments" carried out between 1920 and 1975; but fellow parapsychologists as well as nonbelievers found much to criticize about his selection (Beloff et al. 1980). In another article published (by coincidence!) in the same journal issue as Beloff's, Robert Jahn (1980) cited as the strongest evidence for taking *psi* seriously a different set of results than those offered by Beloff. Still another set was offered by others on a different occasion again (De Beauregard et al. 1980).

  An attempt to make inadequate data convincing is meta-analysis. This statistical technique seeks to assess the validity of a set of studies none of which in itself may have given a statistically significant result. If, for example, this came about because all the samples examined were too small, then some means of aggregating them to form a larger sample might give a more conclusive answer.

However, no amount of aggregating data can yield a valid result if the problem with the individual studies is inherent in their protocol, say, inadequate controls; and often there is insufficient information about the individual studies being aggregated to decide whether or not aggregation is in fact warranted. Meta-analysis in this context is just another instance of overinterpreting inadequate data; it's resorted to because of the lack of reliable means for producing the phenomena one wishes to study. Thus meta-analysis is controversial at best (Mann 1990).

- That episodic apparent success demonstrates the reality of the claimed phenomenon.

  But really, the fact that something anomalous happens not just once but occasionally does not begin to prove that it's a regular phenomenon rather than a sporadic glitch.

- That because something is not known, or because science has no explanation, therefore an anomalous claim should be believed.

  The fact that we lack scientific explanations for many things doesn't establish that there exist forces or things of any particular identifiable sort; that card-guessing experiments have given some results apparently outside those expected by chance doesn't establish that there exists a sixth sense.

- Selecting confirming data and ignoring the rest is a commonly committed fallacy, not necessarily with deliberate evil intent.

  When Velikovsky was scanning books on physics and chemistry and then citing snippets that seemed to support his view, he was not, I believe, consciously trying to hoodwink anybody; he fell into the common trap of selecting just the evidence that suited his purpose, instead of weighing all the possibly pertinent evidence pro and con.

- The purportedly confirming scientific data cited by true believers have often been superseded.

  Outsiders to scientific research, even erudite scholars like Velikovsky, don't realize that citing from a thirty-year-old physics text, say, carries no weight at all when on many points physics has gone considerably further in the meantime.

Debunkers

The disbelievers for their part are prone to commit certain characteristic fallacies of their own:

- Science is taken to be the supreme arbiter: if the contemporary state of scientific knowledge leaves no apparent room for the anomaly, then the anomaly must be spurious.

  In reality, of course, history recounts a succession of revolutionary scientific discoveries that, in the light of earlier scientific knowledge, had seemed impossible and instances of so-called pseudoscience that turned out to be substantive.
- Related to unreasonable reliance on "science" is the implication that there are experts who have the appropriate expertise to pronounce on an anomalous claim. In point of fact, in the most knotty and substantive anomalies it is by no means clear which discipline or which expertise is the most germane.
- While the believers often cite outdated bits of scientific knowledge or theory, the disbelievers often rely on outdated notions about the nature of science.

  Frequently an unorthodoxy is called pseudoscience because it fails to use the scientific (or hypothetico-deductive) method. But not for decades has philosophy of science regarded that scientific method as the hallmark and criterion of science (Bauer 1992a).

  Almost as common is the assertion that to be scientific, claims must be falsifiable. This view was actually pushed for a brief time only, again decades ago, and it has few if any remaining supporters among philosophers. As to philosophy of science, debunkers illustrate well the aphorism that having merely a little learning can be a dangerous thing.
- The debunkers' rendition of the fallacy, that there are only two sides to the issue, is that the believers all believe the same thing, which but a little acquaintance with the literature would show to be false.
- "Skeptics" suggest that converting others to their own opinion is the same as educating them; or that convincing others to disbelieve is the same as helping others to become skeptical.
- Debunkers lump certain things together, without justification. They imply

  —that being illogical or wrong about one thing leads to being wrong about others; whereas in point of fact human beings are marvelously good at holding mutually incompatible beliefs;

  —and that therefore belief in anomalies threatens the rational conduct of democratic society;

  —and that if one test demonstrates a claim of dowsing or extrasensory perception to be false, therefore there is nothing at all to dowsing as such or to extrasensory perception as such.

Yet if such claimed powers *do* exist, one should hardly be surprised if some people were better at it than others, more successful at some times than at others, and perhaps only under some quite special circumstances—succeeding perhaps as rarely as an athlete might broad-jump more than twenty-nine feet.

—That since there are hoaxes or frauds, the whole subject is fraudulent.

That mediums have been found to cheat is true enough, but not all mediums have been shown to cheat all the time. Admittedly the persistent occurrence of fraud lessens the credibility of phenomena that cannot be demonstrated at will. Yet we know that hoaxers and confidence tricksters will flock to situations that offer the best opportunities to cheat without being detected. If *psi* is a real but rare and capriciously episodic phenomenon, then hoaxing and fraud are almost bound to become common around it.

Explanations That Don't Explain

It often seems as though enthusiasts are satisfied with so-called explanations that take one no nearer to actual understanding, for example: "The Eastern religious philosophies are concerned with timeless mystical knowledge which lies beyond reasoning and cannot be adequately expressed in words. The relation of this knowledge to modern physics is but one of its many aspects and like all the others, it cannot be demonstrated conclusively but has to be experienced in a direct intuitive way" (Capra, in Bernstein 1982:336).

As Jeremy Bernstein aptly asks, "If this is so . . . then what is Capra doing writing a three-hundred-page book about the relation?" (Bernstein 1982:336). One might also ask, what exactly is meant by "mystical knowledge"? If one knows something mystically, is it then beyond doubt true? What sorts of things can be known mystically? Any that have tangible physical consequences? If so, could not those consequences be tested nonmystically? And what actually lies "beyond reasoning"? How does one go about experiencing in a "direct intuitive way"? Can one learn to do so? If so, what does one actually do? What does it feel like to know something mystically? And those are by no means the only questions to be answered before that passage could be said to carry useful meaning.

Another typical piece comes from *Supernature* (Watson 1973:4): "So biochemical systems exchange matter with their surroundings all the time, they are open, thermodynamic processes, as opposed to the closed, thermostatic structure of ordinary chemical reactions. . . . This is the secret of life. It means that there is a continuous communication not only between living things and their environment, but among all things living in that environment. An intricate web of interaction connects all life into one vast, self-maintaining system. Each part is related to every other part and we are all part of the whole, part of Supernature."

To some degree the conclusion is noncontroversial, even banal: we all are indeed and everything certainly is part of the universe. The neologism "Supernature" signifies that something more than that is meant; but what exactly? And one might also ask for elucidation of "self-maintaining," "continuous" communication, and so forth. (The point here is that the cited passage lacks meaning. The comments about "closed," "thermostatic" chemical reactions also happen to be technically erroneous.)

In science, explanation means showing how something fits into the corpus of accumulated science. Anomalistics, by contrast, subsists in a sort of vacuum: it stands somehow contrary to what would otherwise be the relevant bits of mainstream science; it lacks obvious connections. So the search for intellectual connection, like that for social connection, often leads to nonscientific sources, to folklore perhaps, or to mystical knowledge, or to other contemporary anomalies. Thus one finds some Atlantists making connections with catastrophism and spiritualism; or some cryptozoologists—to the dismay of other cryptozoologists—suggesting that Bigfoot might be a psychic entity, thereby "explaining" their elusive nature and avoidance of photography through the occurrence of glitches.

Believers, then, may offer connections or explanations that don't really explain. For their part, the nay-sayers are so sure that the anomalous claim is spurious that they are apt to push quite dogmatically the first half-baked or partly baked (Good 1962b) "explanation" that jumps to mind. A bright cloud-like UFO was seen in August 1986, about 10 P.M. in the eastern United States. The UFO-debunkers rushed in with explanations: a meteor shower, a barium cloud, a satellite burn-up, explosion of a Japanese satellite, pieces of dust from a comet, misperceptions by untrained observers, moonlight reflected off a satellite's mirrors. But only one of those could be correct, and it hardly lends credibility to self-styled "skeptics" when they offer a grab bag. In point of fact, careful measurements and plotting of satellite paths showed that this was probably propellant venting from a satellite and reflecting moonlight (Oberg 1986–87). Again, several journalists have claimed, independently and at different times, that their particular published but invented stories created the Nessie myth; but at most only one of those confessions could be true—and since several such confessions are bound to be false, why should one believe even one of them?

Shibboleths

Anomalies accrue a believers' folklore: manifold details that are eminently plausible in the context of belief and that are supposedly well established. These shibboleths deserve to be closely examined, however, for the reliance put upon them is frequently unwarranted.

In ufology, there is a shibboleth relating close encounters with UFOs to strong emissions of energy: the electrical systems of automobiles fail, or the ground becomes scorched or radioactive. While no single such incident has been indubitably proven, the "pattern" exists, and the shibboleth fits the preconception of some believers that UFOs have access to tremendous sources of power that make possible maneuvers and voyages beyond the capabilities of terrestrial technology. In parapsychology, shibboleths clothe *psi* in what would be plausible characteristics of human capabilities (getting tired, needing a supportive environment, and so on). For Loch Ness monsters, shibboleths underscore the animate nature of the phenomenon.

Shibboleths can encumber the believers' quest, misguiding further investigation. Once it's accepted that a decline effect characterizes psychic performance, for example, research can be led astray by taking any set of observations that shows a decline effect as therefore suggestive of the action of *psi.*

Debunkers have their own shibboleths. That psychic effects occur best in the absence of doubters is accepted by them as well as by many believers—but the debunking explanation is that psychic effects are often produced by fraud and that believers are inevitably less on guard against cheating than are nonbelievers. The manner has been described, for instance, in which Uri Geller plays on the need for a psychic-friendly environment to have experimental controls relaxed to an extent that leaves ample opportunity for chicanery. For the naysayers, then, it is a shibboleth that psychic equals fraud or mistake.

Simple criteria proposed for distinguishing between pseudoscience and science turn out to be shibboleths based on notions of what science ought to be rather than on actual characteristics of science. Thus one of Langmuir's postulated characteristics of "pathological science," which I take up in chapter 4, is that rapid growth of publications is followed by a rapid decline, as with biological epidemics (Bennion and Neuton 1976), whereas in real science a new subject shows an exponential growth that then flattens off. Yet this type of S-shaped growth, supposedly characteristic of real science, is only speculated, not measured. In exactly the same connection, another author suggests that proper science shows not a flattening off but a decline (fig. 2).

The latter seems somewhat more plausible. At first, great interest is aroused, then things settle into some sort of routine attended to by a number of spe-

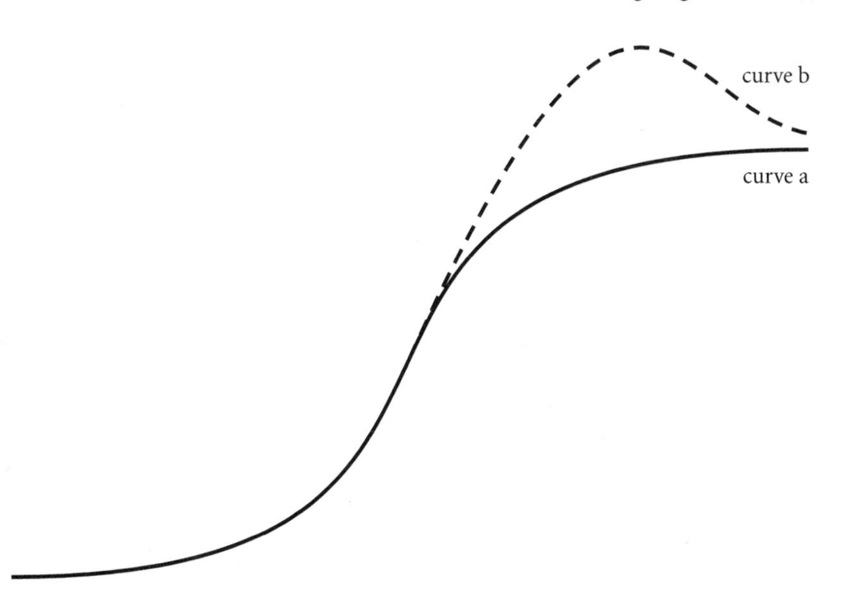

Figure 2. The Growth of Science

How science grows: curve *a* according to Bennion and Neuton (1976); curve *b* according to Franks (1981:128). Both are speculation unsupported by empirical evidence.

cialists that is smaller than the number of people who were initially attracted by a new topic.

But if real science shows a gradual decline of some sort whereas pseudoscience shows a rapid decline, the distinction becomes only a matter of degree and therefore a less useful criterion. Above all, it remains to be demonstrated that topics in mainstream science show at most only a gradual decline in publications. I would not be surprised if high-temperature superconductors, say, showed quite a rapid decline in publication after the initial enormous burst of activity during a few years.

Knowledge seeking is a trial-and-error, boot-strapping sort of thing. Along the way, hunches—working hypotheses that range from staid to fanciful—are tested. Shibboleths can be viewed as working hypotheses that have become entrenched without adequate evidence or have remained entrenched for too long. Mainstream science too, not just anomalistics, is replete with shibboleths. The notion that elementary particles must obey "parity" was held firmly but on quite flimsy grounds, until the Nobel Prize–worthy discovery of the violation of parity (Bernstein 1967:52). The separate conservation of mass and of energy was part of the conventional scientific wisdom until Einstein showed that actually $E = mc^2$. All biology agreed that once cells had differentiated they could not again be made to de-differentiate—even as we now know that the healing of wounds typically or often involves the de-differentiation of mature cells (Becker and Selden 1985). In geology, "the uniformitarian tradition that made geology a science in the 19th century" became a shibboleth in the twentieth so that "Bretz's Spokane flood hypothesis [in the 1930s] appeared as anathema to many of his contemporaries" (Baker 1978:1255).

Some abstract principles are also shibboleths: that open-mindedness is always a virtue, or that interdisciplinary effort is always a good thing. Such generalities are common red herrings in anomalistic controversies, distracting attention from the actual claims.

Much more support and many more allies can be gained by pleading for open-mindedness and fairness than by offering suggestive evidence that Bigfoot exists or that a close encounter with Venus caused the biblically described fall of Jericho's walls. A cadre of social scientists came to Velikovsky's defense as a matter of fair treatment on account of the manner in which he had been denigrated by his critics; it was said that Velikovsky the man was being defended irrespective of the validity or otherwise of his views. But it's fallacious to suggest that the nature of criticism be independent of the substance of a claim; after all, how respectfully should a man be treated who criticizes Establishment science for not publishing his thesis of the inhabited hollow Earth? Still, sarcastic debunking routinely brings to the side of anomalist claimants some moral support from people offended by the sarcasm without necessarily being enamored of the substance of the anomalous claim. Benveniste received such support from scientists as well as from philosophers of science (BBC Television 1994; Schiff 1995); Rupert Sheldrake from some biologists and psychologists (BBC Television 1994), as did Velikovsky and Reich.

Reification: Saying So, Makes It So

Shibboleths like the decline effect illustrate the common fallacy of reification—asserting by implication, by giving it a name, that a thing exists. "The trials showed the decline effect" sounds like the observation of a real phenomenon where a neutral description of what took place might not: "The purported psychic's scores declined as the trials continued."

Amnesia is an undoubted fact; people do sometimes lose memory of certain events. But we do not fully understand the mechanisms that produce it. To talk of "collective amnesia" as Velikovsky did is to name something that may not exist. As to whether such memories can be recovered, that is a question, which is begged by the phrase "recovered memories."

So we can and do give names to things that may not or do not exist. Anomalistics is replete with such terms: ghosts, automatic writing, cross-correspondences, dermo-optical perception, etheric body, ley lines, out-of-body experiences, orgone, and so on. The frequent use of these names in everyday speech and popular literature surely contributes to some level of public belief that the things named actually exist. Ask the woman in the street who Nostradamus was and a likely answer is, "A person who was able to see into the future." Perhaps this is why so many people, when asked, say that they "believe in" angels, the Bermuda Triangle, dowsing, poltergeists. It's not so much that they've thought about it as that they've heard the terms so often.

Popular in the New Age are asserted "others": healing powers are said to work through "other energies" at "other frequencies," perhaps under instructions from spirits in "other dimensions." But those phrases including "other" refer to things for which we have no evidence at all. They are "other" *because* we have never detected or measured them.

A similarly rich assemblage of things that exist by name alone is comprised by phrases starting with "alternative." There are demands for an "alternative" science, less biased against anomalies, using its power to exploit the potentialities inherent in the anomalies. But science is what it is and not anything else. It was not deliberately designed; it evolved as the best practical way of getting knowledge about the tangible world. What reason do we have to imagine that we could consciously devise a better approach? An "alternative science" is science fiction, not any real type of science.

So too with alternative medicine, alternative technology, alternative lifestyles—these are speculations and dreams, not actual practices capable of being adopted without losing some of the characteristics that the activity has without "alternative" in front of it.

But reification is not restricted to anomalistics. As social science attempts to find categories of human behavior, it necessarily devises terms like "anger," "deviance," "intelligence," and so forth that may however have no actual existence as "things." Social science, dealing less in simple, palpable objects than natural science does, is thereby in greater danger of reifying, and is in that respect in much the same boat as anomalistics.

The dilemma seems insuperable: without assigning names, one can't discuss what one is studying; but once a name is given, it easily comes to assume its own reality. Surely "unidentified flying object" is a reasonable term to use temporarily about unidentified things reportedly moving around in the skies; and initially it would not be too difficult to conclude—after investigation and if that is how it turned out—that the objects were not "flying." But after "UFO" has been long enough in common usage, it takes on a life of its own. We forget that the indubitable phenomenon is of people reporting something, not the existence of actually demonstrated objects.

It is exceedingly important, then, to distinguish between a name and the thing to which that name attaches. That doesn't seem to come naturally. Perhaps we're still somehow influenced by the cast of mind of times past, of sympathetic magic, when things and their names were not clearly distinguished even in principle, when speaking the right words could conjure up spirits, when there were taboos against pronouncing some words, such as for the Hebrews the name of God.

Even when we've named things that don't exist, however, the very fact of the naming signifies something about human culture and human knowledge-seeking. Even if every single claimed anomaly were to be spurious, humankind's fascination with these possibilities would still be worthy of study and understanding.

Choosing Names

Even apart from reification, names are a perpetual source of possible confusion in anomalistics. For what is actually the same "fact," the opposing sides use different names, implying different causes and therefore different degrees of plausibility. Hypnosis is medically recognized but mesmerism is pseudoscience—even though Mesmer described or used the phenomenon of hypnosis. Psychosomatic illness, the placebo effect, and spontaneous remission are medically recognized, yet the same events are labeled pseudo when they're called bone pointing, voodoo, witchcraft, faith healing, therapeutic touch, or miraculous cures.

Again, that the Earth is an interlocked system of mutually interacting physical and biological processes could be expressed as, say, A↔B, where A are all the physical processes and B all the biological processes on Earth; but speaking of it as the Gaia concept, where Gaia is the name of a Greek goddess, might imply that the interaction is somehow conscious or purposive.

The term "atavisms" affords a more subtle example: "Factors deep in the unconscious mind which . . . are survivals of man's pre-human ancestry" (Cavendish 1989:45). Spare (Cavendish 1989:224–27) believed they were *things* that he could conjure up in visible form, which makes them distinctly different from what Jung meant by the same term (Jung's synonym was "archetype") or what Freud had in mind (the "id"). When we talk of "the archaic portions of the human brain," we may be referring to one or another of those, or to something else again.

Ignorance

Ignorance of various sorts plays an often crucial role in arguments over anomalies.

The world is full of things for which science has no explanation. According to the conventional scientific wisdom, explanation will come in due course; according to some anomalist believers, science has gone astray and an alternative science is needed to explain those things. That difference in presumption colors much of the arguing. Scientists and debunkers of anomalistics do not emphasize science's ignorance. They are not particularly comfortable about it and certainly don't want it as prominently displayed as science's knowledge; they are reluctant to admit the depth of science's ignorance. Yet perhaps the most apposite assessment of the gaps in modern scientific knowledge is Lewis Thomas's remark that the greatest of all the accomplishments of twentieth-century science has been the discovery of human ignorance (Thomas 1981:49). Modern science has revealed to us a tremendous range of things we do not now understand.

There is also much ignorance about the history of ideas and the history of science, and, even more to the point here, there is widespread ignorance of the history of anomalistics and of these specific associated points:

- that there have been some major vindications of once denigrated unorthodoxies;
- that there tend to be recurrent claims of probably spurious anomalies;
- that anomalistics is not properly the subject of "science" (natural science) but rather is often interdisciplinary among the social and natural sciences.

In addition to unawareness of these general points, there is much specific ignorance in arguments over unorthodoxies, about details having to do with each particular topic. Many debunkers are surprisingly uninformed about the nitty-gritty of the anomalies they denigrate—not to mention the occasional deliberate avoidance of becoming informed because it would be a waste of time. Nessie buffs get weary of would-be debunkers who don't know how vast Loch Ness is, or that dozens of independent groups over several decades have consistently noted sonar echoes from apparently large and moving underwater objects, or that the authentic 1960 Dinsdale film was vetted by experts from Britain's Joint Aerial Intelligence Reconnaissance Centre, and much more.

Matters that the technical experts may be aware of can be entirely unknown to the general public, the media, and anomalist investigators. "Angel hair" supposedly floating down from UFOs turned out to be spider webs—but who apart from arachnidologists knew that some spiders use wind currents to migrate over long distances at high altitudes?

The most consequential ignorance encountered in anomalistics is not, however, that displayed by individual participants in the controversies but rather the substantive illustrations of Lewis Thomas's aphorism. Investigations of reported anomalies continually reveal previously unremarked gaps in our knowledge. For example, claims that cattle have been found mysteriously mutilated "with surgical precision" revealed our ignorance of how many cattle die naturally over the course of a winter and of what carcasses look like after being naturally scavenged by coyotes or birds. A pertinent cryptozoological quest was for the *ri,* which, New Guinea natives knew, was distinct from the dugongs. A couple of cryptozoologic expeditions observed specimens, which indeed behaved differently than dugongs, spending on average about ten minutes underwater at depths of between thirty and fifty feet, whereas dugongs were believed to submerge for only about a minute at a time. Eventually, closer inspection revealed that the *ri* are nevertheless dugongs, and our ignorance of dugong behavior decreased.

The world is full of natural phenomena of which most of us are quite ignorant, and many of which science is ignorant. Research in ufology and parapsychology and cryptozoology often encounters these specific details of nature that are at the same time entirely "natural" but also unsuspected and thereby mysterious.

Closure

Science progresses as congruent knowledge is added, as apparently incongruent data are found actually to fit, as smaller and larger disagreements continually get settled. However, outside the natural sciences, in the humanities and the social sciences, some disagreements persist for centuries and seem likely never to get settled: over ontology (what things exist in the world?), or over epistemology (by what means can true knowledge be obtained?), or over the proper degree of realism or relativism or constructivism (to what degree does our knowledge of the world correspond with the real world out there?). In this, anomalistics may be more like natural science. It does not suffer inherently the fundamental ideological divisions that the humanities and social sciences cannot avoid; and even some very long-standing questions—"Does *psi* exist? Are there paranormal powers?"—seem as though they could eventually be settled one way or the other.

That this last question has remained unanswered despite more than a century of serious study indicates, some would hold, that such powers do not in fact exist; why suspend judgment any longer? But knowledge can only be gained in its own good time, up to which time even genuine phenomena can remain frustratingly capricious and irreproducible.

Closure can come only when debunkers and believers agree, and that will happen only when the mysteries are thoroughly understood.

Fraud

Anomalistics lacks the customs that discourage fraud in mainstream science and scholarship: the instilling of norms of ethical conduct in future researchers and the peer review that filters knowledge claims. Detecting fraud committed by the likes of fake psychics is also hampered by the inherent uncertainties of the purported knowledge. So anomalistics is not only blessed with great intellectual riddles but also plagued by charlatans and hoaxers, attracted to fields in which their antics may not be readily discoverable.

Some confidence tricksters may have started as, or continue to be, genuine believers: perpetual-motion buffs who just haven't found the right trick yet (Ord-Hume 1977:9–10), or at Loch Ness, individuals like Frank Searle (Bauer 1986). They may also have been themselves deceived by nature's tricks: the fact that many alloys of gold are themselves golden in color though still containing unconverted base metal may have convinced some alchemists that they were succeeding in growing gold by conversion of that base metal because the amount of gold-colored material increased. If there are genuine physical mediums, on those occasions when the *real* phenomenon unaccountably refused to cooperate, no doubt they would be tempted to cheat to avoid disappointing their customers.

Fraud can send serious investigators in a wrong direction, and it certainly serves to further discredit, in the public eye and that of mainstream scholarship, studies that enjoy little enough credibility to begin with. But even as hoaxes are a common feature of anomalistics, still the debunkers' usual stance is not justifiable, that the occurrence of hoaxes necessarily or of itself disproves the claims.

Actual Cases

The last two chapters have emphasized general points on which anomalistics shows distinct differences from the established sciences, albeit those differences sometimes turn out to be only matters of degree. But generalities alone cannot give a realistic feel for the difficulty of reaching an informed opinion about controversial and unorthodox matters. One needs also to enquire into actual cases.

The following chapters will summarize some notorious knowledge fights:

Chapter 4—Some claims by competent mainstream scientists that are commonly labeled pseudoscience or "pathological science": N rays, polywater, cold fusion.

Chapter 5—Cases from the borderland between science and anomalistics. For example, over the centuries, interactions between life processes and electromagnetic energies have sometimes and in some places been "scientific" and at other times and in other places "pseudoscience."

Chapter 6—Claims made by nonscientists that are commonly called "pseudo": Immanuel Velikovsky's and Wilhelm Reich's propositions; Loch Ness monsters; extrasensory perception.

In each case I give the briefest of synopses of the main issues for readers not familiar with the particular controversy, with references to fuller treatments. These synopses are not offered as any sort of balanced accounts; they are intended only to bring out points that seem salient from the viewpoint of comparing the practices of science and of anomalistics.

4 Pseudoscience within Science

Occasionally the public learns that some truly portentous scientific breakthrough has turned out to be a dead end. That's thought to be uncharacteristic of science; if enough publicity surrounded the initial claims, the reaction is often strong, and some will cry "pseudoscience."

But it happens all the time in science that intriguing phenomena turn out to be spurious, and marvelous ideas happen not to pan out. Going wrong, occasionally in major ways, is part and parcel of science, of humankind's venturesome exploration of unknown intellectual and physical territory. Frontier science suffers missteps galore.

This is too little understood. Discussions of error in science seem colored by a need to explain how the erring ones erred and at least implicitly how or why they should not have. Elsewhere (Bauer and Huyghe 1999) I have argued that there are no practical safeguards to check even highly accomplished scientists from sometimes committing what some of their peers will later call pseudoscience. Here I illustrate that for some well-known instances: N rays, polywater, and cold fusion. For comparison, I mention the case of high-temperature superconductors, where the behavior of scientists was very similar to that of those in cold fusion while the punditry was very different.

Intellectually, the scientific community is composed of more-or-less overlapping disciplines, which in their turn are overlappings of subspecialties, with the actual working communities often of the order of scores or hundreds of people. Normal science is carried on by these "invisible colleges"—"invisible" because they are linked intellectually but not physically.

N rays

"[T]he history of science abounds . . . with examples of *collective* errors that have given rise to entire schools of false thought. . . . [O]f all such infectious illusions there is none to compare with the remarkable story of N-rays" (Rostand 1960:12). This is perhaps *the* canonical case of *mainstream* pseudoscience, or "pathological" science.

René Blondlot, at the University of Nancy (hence "N" rays), announced his discovery in 1903: a new form of radiation, emitted by both living and inanimate bodies, able to penetrate aluminum but not lead, able to be refracted by aluminum prisms as light is refracted by glass.

For several years N rays were studied by scores of scientists in France and hundreds of papers were published. Yet scientists in other countries were never able to reproduce the radiation. An American physicist, Robert Wood, observed the experiments in Blondlot's lab: in darkness, visual observation was used to detect on measuring scales the spots of light that N rays produced. Surreptitiously in the darkened room, Wood removed the aluminum prism. The measurements continued to be read out as before. Evidently optical illusion was causing spots of light to be imagined at expected values along the scales.

This demonstration convinced almost all the scientific community that N rays do not exist, but Blondlot and a few others nevertheless persisted in their belief that N rays were real.

Connectedness

Blondlot was a member of the French scientific Establishment, a highly respected physicist who had been awarded several prizes. He had been particularly praised for showing that X rays moved at the speed of light.

The *method* Blondlot had used for measuring the speed of X rays depended on variations in visually observed intensity of electric sparks. However, this same approach of visual observation caused him to fall into error in the case of N rays.

The *facts* Blondlot reported were confirmed by a number of his fellow scientists, not only in his laboratory but also elsewhere in France, which gave Blondlot good reason to think his discovery a genuine one.

The *theory,* that new types of radiation awaited discovery, was at the time entirely in the mainstream. X rays and radioactivity had been discovered just a decade earlier, and some years before that Hertz had discovered radio waves.

There are neither social nor intellectual grounds, then, for describing the N-ray episode as outside the usual practices of the science of its time. It may not have represented the best possible practice, but neither was it so egregiously disconnected as to warrant being labeled "pseudo" or "pathological." It just happened to be a false trail at the frontier of science. The label of "pathological" in such cases is an instance of *odium scholasticum;* orthodoxy is irate and overreacts because the erroneous practice is so *similar* to standard practice, not because it is egregiously different.

Illustrative Points

Not Fraud — On occasion, Blondlot published major revisions of some of his studies, for example, assigning N rays to the ultraviolet rather than, as at first, to the infrared. Self-deception was at work, not any attempt at fakery.

Being Led Further into Error — About a year after the first announcement of N rays, Blondlot reported yet another type of radiation, N' rays, which had exactly opposite effects to N rays. To people not already convinced of the reality of N rays, this might have aroused suspicions that something was not kosher, that a reexamination of N rays was called for rather than ascribing a greater difficulty in observing N rays to yet another unknown form of radiation.

It may not be unfair to compare the psychology of Blondlot's approach here to that of Wilhelm Reich when he ascribed a decreased effectiveness of the beneficial orgone energy to the presence of its opposite, "deadly orgone energy" (DOR).

Cheap Shots — The French Establishment treated Blondlot very considerately even after most people had concluded that N rays were fictitious. Scientists in other countries were not so kind: "The Belgian physiologist Augustin Waller remarked that the rays of Nancy should be called 'rays of suggestion,' alluding to the well-known studies on suggestion by the psychiatrists of the Nancy Medical Faculty" (Nye 1980:141).

Ethics — It's generally agreed that lying, cheating, and deceiving are inappropriate conduct, and that when they occur they discredit the perpetrator and his cause. Admirable ends do not justify despicable means. Science is supposed to be particularly observant of ethical niceties.

However, as in the particular applications of any principle, we tend to be more lenient when the cause seems to be a really good one. Robert Wood deliberately deceived Blondlot and his coworkers by surreptitiously interfering with their experiments, yet he is applauded rather than criticized for it. It may be pertinent that Wood was not only a physicist but also "the exposer of countless frauds of spiritualist mediums and other perpetrations" (Price 1975:156).

Context

What grounds then are there for taking N rays as exemplary of pseudoscience or of pathological science?

Those terms are much more pejorative than just the neutral but fully descriptive "scientific error." Perhaps Blondlot can be justly criticized for not recognizing the force of Wood's demonstration, but that was after the damage had already been done, the major error committed. Again, Blondlot did have strong motives for wishing N rays to be real, moreover not purely personal motives—but then every scientist wishes to make discoveries.

The historian Mary Jo Nye (1980) points out that one reason for the excitement aroused by N rays was their emission by living as well as inanimate bodies, raising the possibility that they could be a clue to a connection between spiritualist or psychiatric phenomena on the one hand and material phenomena on the other, at a time of "renewed scientific interest in psychiatry and spiritualism" (145). It had been not Blondlot, however, but a medical researcher, Auguste Charpentier, whose papers "created a sensation" because of the apparent origin of N rays in the nervous system; and it was the eminent physicist and physiologist d'Arsonval (the d'Arsonval galvanometer was a standard instrument in pre-digital times) who found that the emission of N rays increased with increasing intellectual effort.

Motivation for wishing a grand discovery included that French science was seeing its reputation decline while that of German science was rising. Also, France was decentralizing its scientific program and Nancy was one of the provincial universities selected to lead this effort at renewal.

The god-professor was a characteristic feature not only of French science but also in Germany; and British universities too had a tradition, lasting to the middle of the twentieth century, of a single full professor appointed for life as all-powerful head of a department. People who attained such a position, like Blondlot, could find it quite difficult to recognize that they had made an error, let alone admit it publicly.

Most historians recognize that episodes like that involving N rays are not properly occasions to heap blame on those who deceived themselves or who were fooled by fate; rather they are occasions to learn something about the pitfalls of knowledge seeking. "The curious error of N rays is much more a sort of mass hallucination, *proceeding from an entirely reasonable beginning*. . . . It mushroomed into a complex that could have been possible only in that short and glorious epoch when physics had suddenly found the first great massive breakthrough in its modern history" (Price 1975:159; emphasis added).

Polywater

Polywater rivals N rays as a canonical example of error in the history of mainstream science. The polywater episode of the late 1960s effectively ran its course within three or four years, as N rays had at the beginning of the century. A book-length study has been published by a scientist with appropriate technical expertise, Felix Franks (1981).

The discontinuity at a physical boundary has remarkable effects: for example, some substances will spontaneously spread themselves along a surface into a layer exactly one molecule thick; surface tension permits insects to walk on water; mercury rolls about as almost perfect spheres of liquid.

In Russia, surface science had long been intensively studied, including how the properties of water are affected by surfaces. In the early 1960s, Nikolai Fedyakin observed that water sealed in narrow capillaries slowly formed a second column above the original one. The water in that new column didn't freeze or boil like ordinary water. In the late 1960s, the internationally respected Boris Deryagin (Derjaguin) brought this phenomenon to wide attention as a new form of water, at first called "anomalous water," which had a 40 percent higher density than ordinary water and characteristically different refractive index and vapor pressure as well as freezing and boiling points.

Connectedness

As with N rays, the key observations followed quite naturally from a normally proceeding mainstream research program. The *fact* that water in confined spaces and at surfaces has different properties than bulk water neither was nor is in dispute, though in the 1960s it was only in Russia that organized studies were under way (Franks 1981:27–28). The *methods* used were quite standard. As to *theory,* there was no lack of speculation based sufficiently on accepted notions to be published in respectable journals.

Though the international community had high respect for a few prominent Soviet scientists, much Soviet science was published only in Russian-language journals and effectively outside the international mainstream; the latter became aware of "anomalous water" only after Deryagin took up the matter. Then, however, so renowned a scientist as J. D. Bernal would venture that anomalous water "is the most important physical-chemical discovery of this century" (Franks 1981:49).

It seems that American researchers were alerted by reports from the London office of the U.S. Office of Naval Research (whose observers were viewed by British scientists as somewhat in the nature of industrial spies; see Franks

1981:56–57). The polywater furor really exploded after the well-known American spectroscopist Ellis Lippincott reported the infrared spectrum of anomalous water as clearly different from that of ordinary water. He called it polywater rather than anomalous water because in his view it was a polymer of water—a number of water molecules joined together. Polywater was discussed at several of the prestigious Gordon Research Conferences (Franks 1981:124).

Felix Franks speculates that international competition may have been one factor in the intense burst of activity; American science and government were still smarting from the Soviets having been first into space in 1957 with the Sputnik satellite (Franks 1981:58).

Illustrative Points

Immediate Ill-Informed Dismissal — As typically in anomalistics, preconception more than evidence colored individual opinion: "The skeptics were accusing the believers of naivete and undue gullibility, and the believers were accusing the skeptics of arrogance and lack of imagination" (Franks 1981:139).

Pundits often seem unable to appreciate just how technically difficult much of science is. As to polywater, "practical problems posed by this type of research are formidable. . . . The progress in experimental studies of water is dotted with claims of anomalies" (Franks 1981:46–47).

Some people dismissed polywater as impossible, because if it really were more stable than ordinary water, then all the water on Earth would long ago have turned into polywater: "Astonishingly few of the polywater scientists considered the implications of thermodynamics, even though they lead to an immediate negative conclusion. . . . Since polywater has a lower vapor pressure than ordinary water, with which it is in equilibrium through the vapor, it follows from the first and second laws of thermodynamics that the polymer must be the more stable form. . . . [E]ither . . . billions of years of water waves and water rains washing over silica beaches had not produced the stable form of water [polywater had allegedly been observed in tiny capillaries of silica] or . . . the laws of thermodynamics are wrong" (Eisenberg 1981:1104).

That argument had in fact already been made during the heyday of the fascination with polywater. Why had it not won the day then? Because it is not a sound argument.

Though the same reasoning continues to be repeated by apparently qualified scientists, it is just as thermodynamically illiterate as the argument that continues to be made by some "scientific creationists" that the Second Law of Thermodynamics makes biological evolution impossible. Many substances *do* exist concurrently in several forms of differing stability without interconverting even as they are "in equilibrium through the vapor," for example, pure carbon (diamond, graphite, bucky-balls), or phosphorus (red, yellow, black), or selenium (gray, red), or sulfur (monoclinic crystals or orthorhombic ones).

The same argument as this one against polywater would insist that the world should have no carbon in the form of graphite because diamond is a more stable form of carbon. But for a less stable form to convert into a more stable one, there must be some feasible mechanism through which the change can take place, and there are innumerable instances where ordinary conditions do not provide a mechanism.

Even eminent scientists fall into this trap. The Nobel laureate Richard Feynman once remarked (Eisenberg 1981) that there could be no such thing as polywater because if there were, there would also be an animal that didn't need to eat food. It would just drink water and excrete polywater, using the energy difference to maintain its metabolism. By the same reasoning we should have organisms that imbibe carbon in the form of graphite (from the ashes of forest fires, say) and excrete diamonds.

Even beyond that, naive dogmatists ought to recognize that the laws of thermodynamics are empirical laws. That we have not hitherto found cases where those laws break down doesn't mean that we shall never come across such an instance. For that matter, relativity *did* break a thermodynamic law, the law of conservation of energy, the belief that while energy might be converted from one form into another (mechanical to heat, say), it could not be created or destroyed. Along came $E = mc^2$, and it was realized that energy *can* be created or destroyed (or, that mass is really a form of energy).

Then too, it isn't always easy to apply *properly* even valid laws to a given situation; the application of general laws can founder on incomplete specific knowledge. A widely held shibboleth based on the laws of thermodynamics is that perpetual-motion machines are impossible—one can't get energy out of nothing, is the popular slogan. But the Earth is being continually fed energy in the form of radiation and particulate matter from the rest of the universe; perpetual motion on Earth would merely require finding a means of harnessing such incoming energy or matter in some appropriate device.

It may not even be true that we cannot get energy out of nothing. Quantum mechanics seems to imply that the vacuum, long believed to be the very epitome of "nothing," isn't really "nothing," or at least is "nothing" only on average, and that the fluctuations of "something" energy in that "nothingness" could perhaps be harnessed to provide usable energy to our world (Forward 1996).

During the height of the polywater controversy, the thermodynamic argument was properly ignored. After the event, however, few if any people have an interest in defending the erstwhile proponents of polywater, and so the thermodynamic argument is allowed to stand as though it were indeed decisive. But it's no more decisive after the event than it was before.

Apparent Confirmations — As with N rays, the original discoverers had good reason to trust their work since others had reported confirmations.

Cheap Shots — The contemptuous attitude that thermodynamics should have warned competent scientists against polywater is just one of the unwarranted criticisms made of the well-intentioned knowledge seekers who got caught up in trying to understand anomalous water. As Franks points out, "Nowadays any reference to polywater is always tinged with ridicule, but . . . [at the time] many competent and experienced scientists were quite convinced of its reality" (Franks 1981:ix). "Time and again the proponents of polywater were 'accused' of practices that are in fact common and acceptable to professional scientists, such as the publication of ideas and hypotheses not yet thoroughly tested and confirmed—a fairly common practice, as illustrated in *The Double Helix*" (Franks 1981:189).

Role of the Media — Deryagin himself had not intended to make public announcements until further work had been done, but others alerted the media (Franks 1981:52).

A curious point is that a more stable form of water ("ice-nine") that thereby threatened worldwide disaster (all water becoming ice-nine) had been imagined some ten years before the polywater episode, in a science-fiction book, *Cat's Cradle* by Kurt Vonnegut. "A warning that polywater, like ice-nine, might be the most dangerous material on earth alerted the news media, and from then on much of the scientific debate was carried on in the pages of newspapers" (Franks 1981:3).

Who Is Authoritative? — "Water" is studied by a variety of different specialist groups who don't normally communicate intensively with one another, for example microbiologists and cloud physicists. Consequently, important knowledge is not immediately available to those who could benefit from it, for instance, that "certain bacterial spores found on rotting leaves during the autumn provide much better nuclei for the seeding of clouds (rainmaking) than anything previously tried" (Franks 1981:27).

Polywater touched on many subspecialties in thermodynamics, surface chemistry, spectroscopy, theoretical chemistry, and analytical chemistry, among others. The ultimate resolution may have been owing to the expertise of analytical chemists applied to what had been discovered and worked on primarily by surface scientists.

During the public fuss, there was no way for the media to distinguish genuine workers from publicity-seeking but not really qualified pundits (Franks 1981:61), something more usual in anomalistic controversies than in scientific ones.

Patronage — Early research into polywater was actively supported by defense interests (Franks 1981:2). The military (as also technology in general, including medicine) is concerned only with whether or not something works in practice; that there may be no explanation for it, or that theory insists that it should not work, carries little weight. As a result, some very farfetched research has been supported by defense agencies: studies of psychic remote viewing in both the USSR and the United States, "Star Wars" projects like X-ray lasers, and controlled nuclear fusion in Argentina decades ago. In the United States, the Defense Advanced Research Projects Agency (DARPA, earlier ARPA) is known to the scientific community as offering opportunities to have research funded that would be too adventurous for the National Science Foundation to support. It was to DARPA that Fleischmann and Pons proposed studies of cold fusion, and DARPA that had already funded somewhat similar studies.

Ignorance — Water, so abundant and so much studied, actually remains less well understood than such exotic substances as liquid nitrogen (Franks 1981:chapter 1). The physical and electrical shapes of the water molecule lead to unique interactions with other molecules and to water's extraordinarily high boiling and melting points, surface tension, and heat capacity. Unlike any other substance, the solid floats on the liquid (if ice did *not* float on water, then lakes and oceans would freeze completely in winter, not just at the surface, and life as we know it might never have evolved on Earth).

More than half of human tissue is water, but much of that water, is bound in some not fully understood manner. Some of it is "unfreezable"—as is water in tiny mineral pores, for example. Biological interactions are so sensitive to the properties of water that "heavy water," in which hydrogen is replaced by its chemically very similar isotope deuterium, is actually toxic (Franks 1981:11).

Polywater, it finally turned out, was a solution in water of various impurities—yet not the impurities that one would have expected if the water were leaching them out of the quartz capillaries in which it had been prepared (Franks 1981:101). The observations made by Fedyakin, Deryagin, and others were sound observations of a novel phenomenon that the scientific knowledge of the time could not adequately explain.

Recurrence — Not everyone agreed or realized that polywater had been conclusively shown to be spurious. Franks notes, "In 1977 Maryna Prigogine, the wife of a Nobel laureate and a student at the Free University of Brussels, reopened the whole issue in her 200-page Ph.D. thesis" (Franks 1981:141).

The notion that water can be somehow "structured" by contact with something and then retain that structure ("water memory") has again been invoked in recent years by Jacques Benveniste in attempting to explain how biological activity could be shown at homeopathic ("infinite") dilutions of active agents (Davenas et al. 1988). Curiously enough, Deryagin once mentioned Russian pharmacological work similar to Benveniste's: "very very small concentrations dissolved in water of some compounds . . . one molecule per cubic millimeter or even more dilute concentration" had been reported to affect, for example, the heart of a frog (Franks 1981:50). Moreover a half dozen years after the end of the polywater episode, "Soviet scientists have known for some time that fresh meltwater has the capacity to stimulate some biological processes. It has been theorized that the meltwater retains some of the order that is characteristic of frozen water" (Franks 1981:Epilogue).

Ethics — Scientific publication is supposed to be judiciously refereed following careful checking by those who seek to publish. In the polywater episode, rushed publication was standard. Not only the advocates of polywater rushed experiments and publication; "the same was true for the experiments of Rousseau" (Franks 1981:159), who showed that polywater was impure water. But Rousseau has not been criticized for rushing to publish, presumably only because he was on what is now supposed to be the right side.

Grains of Truth — Some genuinely new scientific knowledge was gained as a result of polywater studies. Violations of the vapor-pressure equation for capillaries had been remarked by a number of other workers before Fedyakin and had now been resolved as partly due to dissolved substances as well as the surface forces. Further, points on which science remained sadly ignorant had been identified, such as the fact that "the dissolving power of water formed by condensation of vapour is many times greater than that of liquid water at the same temperature" (Derjaguin 1983:10). Franks writes that "several of the questions . . . raised have not yet received satisfactory answers . . . [namely] that water *vapor* reacts with quartz more readily than does liquid water. . . . Is water adsorbed from the vapor phase onto silicate surfaces a much better solvent than bulk water? Is it more acidic than bulk water? . . . [There remains] the vexing question of the modification (albeit not permanent) of water in the proximity of surfaces . . . [and] the role played by water in the maintenance of the so-called native states of biological macromolecules . . . [and] questions associated with 'bound water'" (Franks 1981:145–46).

Context

What grounds are there for denigrating polywater as pseudoscience?

The major flaw had been a failure to detect contaminants; and it is unwarranted to criticize people who could not quickly do that when studying novel systems. There's no way to ensure absolute purity. Every new combination of materials offers its own challenges, and every new investigation may need some greater level of purity than had ever before been required. Deryagin had taken all precautions possible at the time, including preparing the anomalous water in quartz rather than glass capillaries, it being known that impurities—in tiny amounts—could leach out of glass into water even at ordinary temperatures (Franks 1981:31). The first major American publication made a point of the lack of spectroscopic evidence in the spectra of any contamination (Franks 1981:71). Though it's now widely agreed that polywater is impure water, what exactly the impurities were is not agreed.

There continues to be no dispute that water in confined spaces has very different properties than free bulk water. The polywater claim is that water *removed* from those spaces might keep some or all of those properties; whereas the conventional view is that those special properties are direct results of forces exerted by the confining spaces. Since the forces between water molecules are uniquely strong, it doesn't seem entirely beyond the realm of possibility that water could, perhaps only in some statistical fashion, retain some properties derived from having earlier been in contact with some surface. It's analogous to making a cast, the difference being that water is fluid and therefore we don't expect it to retain any particular shape. But then glass too is a fluid, yet it retains shapes because it is very viscous and slow-flowing at ordinary temperatures. In the solid state, water—ice, that is—can exist in no fewer than eleven different forms that differ solely in the manner in which the individual molecules are packed and bound together; why shouldn't some of those packing modes still show themselves in the liquid state, albeit only over shorter distances than in the solid and perhaps under certain special circumstances, say, of temperature or agitation? That's the case with "liquid crystals, which while flowing like an ordinary fluid can maintain an ordered structure over macroscopic distances," according to the physics Nobelist Brian Josephson (1977) in defending speculations like those of Benveniste.

"Most scientists rejected polywater out of hand for the wrong reason: because it was different," Franks observes (Franks 1981:viii). "In general, the American, Soviet, and European scientific establishments do not come out of the polywater affair looking good. . . . [Their] collective attitude . . . [being] disbelief, mistrust, derision, and lack of grace" (Franks 1981:138).

Scientific establishments often turn out to have been right not because their members are endowed with the wisdom to see into the future or into the unknown unknown but simply because "they are conservative, and in science conservatism pays off more often than not" (Franks 1981:185). In considering polywater, "open-mindedness . . . was 'carried too far, but not much too far,'" including by theoreticians, who showed that "they can explain anything if they want to badly enough, including why a nonexistent substance is stable" (Franks 1981:155). Once again, calling polywater "pathological" is expressing *odium scholasticum.*

Cold Fusion

For some twenty-five years I worked at electrochemistry. I had spent a sabbatical year (1972–73) at the University of Southampton, where one of the world's foremost electrochemists, Martin Fleischmann, was a professor. In the 1970s I turned to science studies, and by the late 1980s electrochemistry and electrochemists had become only memories for me. Yet I was fascinated, one night in March 1989, to see Fleischmann on "CNN Headline News" as the top story of the day—flanked by Stan Pons, whom I had known slightly at the University of Michigan a couple of decades earlier. Naturally I followed the cold-fusion affair with considerable interest.

In electrolysis of heavy water (deuterium oxide) at palladium electrodes, it was said, far more heat was being generated than could be explained by chemical reactions or by the electrical power applied. It had to be a nuclear process. Apparently deuterium ions underwent nuclear fusion inside the palladium.

This was iconoclastic. According to the conventional wisdom, nuclear fusion requires conditions of temperature and pressure similar to those inside the Sun. Physicists had been trying to achieve this for several decades with enormously expensive machinery. Here was a table-top, test-tube experiment at room temperature claiming to do the same thing, very cheaply indeed (though the cost of palladium, a comparatively rare metal, is not negligible—and skyrocketed during the cold-fusion furor).

For several months, confirmations were announced daily or weekly from many parts of the world, even as most (but by no means all) physicists refused to believe that nuclear fusion could be implicated. But then some confirmations began to be retracted and failures to replicate were announced. Soon work on this topic lost respect even among most chemists and electrochemists. However to the present time some competent people and respectable organizations continue to pursue the quest for cheap energy from such systems.

The book-length studies presently available do not provide satisfactory analyses and interpretations of the cold-fusion affair, and not only because the affair is not yet over. All the books tell much the same story about what happened and when, but little beyond 1989 and 1990. The earliest book was a potboiler rushed into print (Peat 1989), but even the three major debunking books (Close 1991; Huizenga 1992; Taubes 1993) published over the next several years commit significant errors (Bauer 1991, 1992b, 1995c). The best-written book about the matter (Bauer 1992c; Mallove 1991) has only the drawback of the author's unwavering belief that cold fusion is real.

"Cold fusion" soon became the same sort of public icon as "Loch Ness monster," applied to quite disparate and substantively unconnected matters. The Internet offers such examples as "Allaire—developers of *Cold Fusion* [emphasis added], a WWW Database Development tool for Windows NT and Windows 95 which enables the rapid creation of interactive, forms-based Web database applications"; a Web page offers links to "British rock, time travel, *cold fusion* [emphasis added], and a source of a DOS fortran compiler," and another describes "the Pitt at Johnstown physics page; a *cold fusion* [emphasis added] mix of comix, academia, and journalism. Includes special sections on Kurt Busiek and Paul Pope." Fleischmann and Pons are regarded as fair game for lampooning, as in the *Journal of Irreproducible Results* (Abrahams 1992).

Connectedness

Fleischmann and Pons were electrochemists using electrochemical methods but venturing explanations belonging to nuclear physics. Fleischmann was known to be an adventurous thinker, the sort of person—like the astrophysicist Thomas Gold (1999)—whose suggestions are always worth attending to even when they do not work out; his competence was beyond question. So the electrochemical community reacted with astonishment but not automatic dismissal. All over the world electrochemists tried to reproduce the Fleischmann-Pons effect. The Department of Energy held a major conference on the matter in May 1989.

The physics community reacted quite differently, with much immediate criticism that was often egregiously ill-informed. The highly respected *Journal of Electroanalytical Chemistry* was described as an "obscure electrochemistry journal" (Love 1989) and Fleischmann and Pons as "little-known electrochemists" (Pool 1989) displaying "incompetence and perhaps delusion" (Koonin 1989). To call a Fellow of the Royal Society like Fleischmann "obscure" is like so labeling a member of the U.S. National Academy of Sciences. Cold fusion was called pseudoscience (Love 1989), "bad science" (Taubes 1993), and "the scientific fiasco of the century" (Huizenga 1992).

Much was made of the lack of detailed replication of results; but electrochemists appreciate the difficulties of defining experimental details precisely enough to ensure reproducibility, most particularly where electrodes with solid surfaces (by contrast to continually renewed, liquid, typically mercury) are concerned.

As later attempts to reproduce the cold-fusion effect became mostly unsuccessful, a new community came into being, a select group of those who had themselves had sufficient success that they refused to accept the phenomenon as spurious. This new community is drawn both from electrochemistry and from physics as well as other disciplines, in particular engineering. Quite a few scientists, otherwise acknowledged as competent, continue to believe that there is valuable substance of some sort or other in "cold fusion," whatever the exact mechanism may turn out to be (Storms 1996). Until Fleischmann's retirement in 1997, he and Pons worked in a laboratory built and funded by a Japanese venture-capital company. The Sixth International Conference on cold fusion was held at Hokkaido in October 1996, the seventh in Vancouver in March 1998, a Symposium on Cold Fusion and New Energy in New Hampshire in October 1998, and the Eighth International Conference in Italy in May 2000. The Japanese New Energy and Industrial Technology Development Organization spent $100 million on "new hydrogen energy." *Fusion Technology,* published by the American Nuclear Society, regularly carries articles on cold fusion (Mallove 1996).

The contemporary cold-fusion community is not very tightly coupled, however, to the rest of the scientific community. Those who work at cold fusion don't get much in the way of grants from the usual sources of funding for chemistry and physics. Their colleagues tend to look askance at their continued involvement—going so far as wanting to strip the title of "Distinguished Professor" from one such maverick (Mangan 1994). There are rarely sessions on cold fusion at the regular conferences in electrochemistry or nuclear physics. So gauging the connections between the cold-fusion community and the mainstream community is not straightforward.

Similarly with the intellectual connections between cold-fusion studies and mainstream science. The *methods* used by Fleischmann and Pons are largely standard electrochemistry, the chief innovation being the addition of heat-measuring ("calorimetric") devices. One of the aspects of science illustrated by this episode is how very difficult it is to apply even standard techniques from one subspecialty to another; for example, there is continuing controversy over the accuracy and reliability of the calorimetry. Moreover when Fleischmann and Pons tried to add measurements from nuclear physics, to detect possible radiation and newly produced atomic species, their work was less than competent. So in method, cold fusion is partly connected but also partly disconnected from the mainstream, in a manner characteristic of interdisciplinary work when individuals try to do it all themselves instead of enlisting specialists from the several relevant areas.

The *facts* alleged are, at bottom, a purely empirical claim: do this, and the following happens—namely, large amounts of heat, so large that we cannot explain it except by reactions whose energy output approximates that from nuclear reactions. Doubts about the method complicate matters, however; the "observation" of heat depends on calculations that include assumptions. The number of experimental variables that might affect reproducibility is vast. There are, for example, almost innumerable variations in the physical characteristics of the electrodes and in the electrical regimen; and there are all sorts of possible contaminants, conceivably active at levels that might be virtually impossible to detect by other means than their interference with the looked-for effect—by no means an unprecedented possibility (see my mention of Jaroslav Heyrovsky in chapter 2). For the time being, the "facts" of cold fusion are disconnected from the body of accepted, orthodox, scientific facts.

There is a significant distinction between cold-fusion-type facts and polywater-type facts. To reproduce cold fusion means to reproduce a process, which is ephemeral. One cannot afterward reexamine exactly what was done to look for clues as to why it was successful or unsuccessful; one is stuck with whatever descriptions were written at the time, and those can only include what seemed then to be significant variables. By contrast, polywater is a substance, samples of which can be preserved and reexamined much later, when additional insight has been gained.

Anomalistics by and large is like cold fusion more than it is like polywater. The assertion of influences—as in astrology or in *psi*—can hardly be disproved since there are so many conceivable ways in which observations could be hampered or helped by adjusting factors that might affect those powers or influences.

As to *theory,* the situation is far from clear. Cold fusion would contradict only some parts of current theory in nuclear physics, and nuclear physicists do not claim their current theory to be the last word; indeed they're anxious to find clues to developing better ones and have made several suggestions in that direction. Some competent theoreticians, from the nuclear-physics community as well as other fields, have suggested how the metallic palladium lattice might form an environment in which fusion could occur, say, by some sort of "concerted" action analogous to what happens when a laser produces coherent light. Before lasers were demonstrated, no one had believed *that* to be possible either. "Although cold fusion was, in terms of 'ordinary' physics, absurd, it was not obviously so; it contravened no fundamental laws of nature" (Lindley 1990:376).

Illustrative Points

Misconceptions about Science — The claim of cold fusion was denigrated as unworthy of proper scientific procedure: "This is hardly the normal state of science. But this is the world of cold fusion, where everything is distorted" (Derra 1989:33); like polywater and homeopathy, it was pathological science (Rousseau 1992).

As I've already suggested, however, startling claims should be compared not with "normal" science but with potentially revolutionary science. There, it turns out, many of the things are typically to be seen for which Fleischmann and Pons were castigated. The quest for high-temperature superconductivity displayed the same breakneck pace, the same use of the popular media for reporting results before publication in peer-reviewed professional journals (Felt and Nowotny 1992). "Pons . . . and . . . Fleischmann were veritable paragons of model behavior compared with what happened in the high-T_C superconductors" (Roy 1989). With cold fusion, "each hypothesis proposed . . . was tested and dismissed in reasonably short order. . . . [E]ach represented science at its best: a bold hypothesis, an experiment, further experiments, failure to replicate, and down the tubes"; the reason for denigrating these cases was only that the ideas "challenged a prevailing idea instead of supporting one" (Oderwald 1992).

So cold fusion need not be seen as an exemplar of mistake-ridden science; it is actually most instructive about the state of late-twentieth-century science, as detailed in the insightful discussion by Lewenstein (1992). It's absurd to assert, for example, that "self-delusion is very much a hallmark of pseudoscience. . . . It is important in science to be quite honest with oneself and with one's colleagues" (Love 1989:7). It's no easier for human scientists than it is for other human beings to avoid self-delusion and to practice being "quite" honest. Self-delusion is not under conscious control, after all.

Was it silly to start the research? During studies of electrolysis in the late 1960s, Fleischmann had observed anomalies having to do with hydrogen isotopes. Pons independently though later had made similar observations (Dagani 1989). In 1972 Fleischmann had interpreted certain evidence as indicating that deuterons in palladium lattice show less repulsion than expected. It was a far-out idea that this might make fusion possible, but Fleischmann and Pons were quite clear about that. For something like five years they bore the research costs themselves because they knew it was not yet reasonable to apply for research grants. But they got more heat out of their systems than chemical reactions could explain, and so by 1988 they had reached the stage where they felt justified in applying for a grant.

Should Fleischmann and Pons not have made their public announcement when they did? They really had no choice. They had applied to DARPA for research funds. DARPA had asked the advice of Steven Jones, a physicist at Brigham Young University. Jones had told Fleischmann and Pons that he was working along similar lines; he was intending to do some electrochemical experiments similar to theirs, he was planning shortly to give a lecture on his work and then to publish, and he invited them to collaborate with him (Lewenstein 1992).

Fleischmann and Pons quite properly informed their university's administration of this. That administration and its lawyers determined subsequent events, influenced primarily by the university's wish to protect patent rights and to reap prestige. (Compare the University of Nancy's seeking status within French science in the N-ray affair, above.) Already at the first press conference, Fleischmann and Pons said that they would have preferred eighteen months of further work before publication. The press conference was an institutional decision, not a personal one by Fleischmann and Pons.

Should Fleischmann and Pons have concluded that, since attempts by them as well as others to make the phenomenon easily reproducible were not succeeding, therefore their initial successes had been spurious? *When* should they have concluded that? And why? Major discoveries in science have been made through the refusal to accept unsuccessful experiments as decisive: for example, Jacob and Brenner on the way to discovering messenger RNA, "sure of the correctness of our hypothesis . . . started our experiment over and over again" (Grinnell 1996).

It's hardly justifiable to castigate people for breaking "rules" that belong only to an idealized view of basic science formed a century or more ago and even then not reflecting actual practice. Nowadays new claims that have substantial commercial implications are routinely divulged at press conferences, for example, with high-temperature superconductors (Felt and Nowotny 1992) or in molecular biology (Werth 1994). Indeed one of the most outspoken critics of the current scene has suggested that the hothouse competition makes news conferences more appropriate than peer-reviewed publication (Roy 1992).

Certainly the behavior of Fleischmann and Pons was regrettable in a number of ways. Their attempts to make radiation measurements were rushed and sloppy and their subsequent attempt to change or fudge those data is reprehensible. But it's not very different from what most scientists do when they venture into new fields and when they are subjected to vicious attack. Believing themselves to be basically correct, Fleischmann and Pons will have seen themselves as correcting an error rather than as fudging data; and it was very natural to try to avoid admitting error in anything—we all are reluctant to do that in arguments when we're put on the spot.

Apparent Confirmations — Fleischmann and Pons must have been powerfully reinforced to believe the validity of their results by Jones's claim of a certain degree of success himself and his suggestion that they collaborate. Only later did it become clear that Jones's observation of excess neutrons was many orders of magnitude less than that needed to explain the excess heat of cold fusion.

In the weeks following the press conference, numerous replications were reported: "By the end of April, a dozen laboratories from around the world had publicly announced partial confirmations . . . [and] at least 40 articles . . . had been submitted to refereed journals" (Lewenstein 1992:69–70). Other ways of inducing deuterium in palladium to undergo fusion had also been reported to work; for example, an electric discharge between deuterium-loaded palladium rods gave intense neutron emission (Britz 1990). Surely Fleischmann and Pons must have felt their discovery to be thoroughly affirmed.

Workers at Brookhaven National Laboratory reported fusion by "cluster bombardment" of certain targets. The difficulty of tracking down all possible sources of error in such experiments and measurements is illustrated by the fact that it took two years for the Brookhaven group to retract its claim (Amato 1993). Yet at no stage was it subjected to the vicious, personal attacks that had been directed at Fleischmann and Pons.

Who Is Authoritative? — As typically in anomalistics and interdisciplinary science, so with cold fusion it is not obvious that there are any genuine authorities. As already noted, physicists were on the whole dismissive whereas electrochemists were open minded. As late as 1995, the American Chemical Society had a poster session on cold fusion at its national meeting in Anaheim, California—the first one since 1989 (Dagani 1995a,b); that in itself might be grounds for *not* classing cold fusion with alchemy or pathological science.

MIT physicists were angry (Anon. 1994b) when their alumni magazine, *Technology Review,* published a review article supporting cold fusion; yet the article came from a well-qualified source, a chemist recently retired from Los Alamos National Laboratory who had himself done relevant experiments.

Role of the Media — The predominant public view of cold fusion as utterly discredited owes much to the choice by the media of which authorities to cite. Jerry Bishop in the *Wall Street Journal* long continued to report in neutral fashion about cold fusion; indeed, in January 1992 he won the American Institute of Physics "Science-writing Award in Physics and Astronomy," for his coverage of cold fusion (Beaudette 1993); yet sometimes Bishop has been criticized for being neutral rather than denigrating about cold fusion.

Cheap Shots — Among a number of unwarranted criticisms was the remark that cold fusion was another instance of the "Utah effect"—spurious X-ray lasers had been reported from there some years earlier (Pool 1989); compare the snide comment that N rays be called "rays of suggestion."

The same arrogance toward institutions of low research prestige was shown by the physicist David Morrison, who remarked that confirmations of cold fusion had come primarily from southern and eastern Europe and the southern United States whereas failure to replicate had been reported from the elite and reliable institutions of northern and western Europe and the West Coast and Northeast in the USA (*NOVA* 1991).

Just being wrong does not always bring this sort of denigration, not even being very wrong on a major point. Lord Kelvin was dogmatically wrong about the age of the Earth; Lord Rutherford called the notion of useful atomic energy "moonshine." Yet both were and are honored for their life's work despite these blunders, whereas Fleischmann's achievements are as if nullified by his possible error about cold fusion. It seems that if one is wrong more or less in line with the contemporary conventional wisdom, the error is no basis for personal attack; but being wrong *and* against the grain brings personal abuse besides any substantive argument.

Recurrence — Cold fusion had already been claimed much earlier—in 1926 by Paneth and Peters, who reported electrolysis at palladium as producing helium (Britz 1990). A patent on such a process had been filed by Tandberg in 1927 (see the very useful chronology of cold fusion, including such precursors, by Lewenstein and Baur [1991]).

Ignorance — As the cold-fusion story unfolded, it turned out that many relevant things had been known to neither Fleischmann and Pons nor their immediate critics.

For example, as just noted, similar claims had been made as far back as 1926. In addition, Fleischmann and Pons had impressed many people by recounting that in one experiment their electrode had become so hot that it had burned its way through the floor. However, as Ligon has pointed out, "Even the most cursory examination of the literature will reveal that palladium cathodes are prone to explosions. See, for example, Moeller, T., 'Inorganic Chemistry: An Advanced Textbook,' . . . page 411, 1952" (Ligon 1989:2)—and the standard accepted explanation for this phenomenon had nothing to do with nuclear fusion.

Impurities in experiments frequently cause trouble, often because there is no reason to suspect their presence until they cause the trouble. Semiconductor physics in the 1930s did not lead to transistor technology because of the unprecedented and therefore unrecognized purity requirements; researchers often could not repeat their own published work (Chapman 1991).

The detection of tritium or helium was taken to indicate that fusion of deuterium had occurred; but surprises about possible contamination by helium or tritium included the facts that heavy water may contain enough tritium to produce as many as five thousand radioactive disintegrations per minute per mg (Christman 1990) and that the diffusion of helium through glass is affected by the presence of hydrogen (Britz 1990).

Even though most nuclear physicists agreed that fusion could not occur under the mild conditions used by Fleischmann and Pons, it turned out that the theory on which that judgment was based was in error by between six and ten orders of magnitude, a factor of between a million and ten billion (Love 1989; Maddox 1989) (which still leaves the estimate many orders of magnitude away, however, from expectations of measurable room-temperature fusion).

Unorthodox and Unwelcome Associations — Some publications on cold fusion appeared in mainstream journals, especially in the early stages. But increasingly the topic was pushed out of the mainstream.

Specialty journals were founded, for example, *Cold Fusion* (Holden 1994a) and *Infinite Energy: Cold Fusion and New Energy Technology* (Dagani 1996). Within the scientific community, the credibility of a subject is dramatically less—close to zero, in fact—if it is treated in such publications rather than in established scientific journals. Anyone, in fact, might be suspicious of publications that carry little of substance at a high price, as with *Fusion Facts,* a stapled, 24-page monthly newsletter of the Fusion Information Center, Inc., with a price tag of $300 annually. Much the same material was to be found in *New Energy News,* a 20-page monthly newsletter at $60 per year from the Institute for New Energy, printed for it by the same Fusion Information Center, Inc., located in the University of Utah Research Park. There have been other initiatives also to cash in on cold fusion even from the earliest stages; thus Clustron Sciences Corporation was "to capitalize on business opportunities in the emerging field of *new nuclear science,*" with the help of the Brightsen "nucleon cluster model" (Clustron Sciences Corp. 1992), which itself, however, remains unknown to mainstream science.

Those publications further hurt the credibility of cold fusion by including other heresies: "reports on both cold fusion and other enhanced energy devices," including (*Fusion Facts* 1994) "Anti-gravity with present technology" and "the notion that cold fusion, ultra sub-atomic particles, gravity, electricity, and magnetism are each only a consequence of space moving in special interrelated geometric formations." One advertisement was for *Electrogravitics Systems: Reports on a New Propulsion Methodology,* the "T. T. Brown Project of 1954 that resulted in an electrogravitic drive for the B-2 Bomber." Conferences arranged by the Academy for New Energy, whose parent is the International Association for New Science, treated cold fusion together with such things as "Alternative Science," "Geometry of Non-Hertzian Fields and DNA," "New Stable Isotopes" (Academy for New Energy 1997).

Fleischmann and Pons had no way to prevent people from creating or asserting these and similar associations. There were other events that they will hardly have welcomed:

- Promiscuous claims were made of excess heat and tritium production in *light* (ordinary) water as well as heavy, and at other metals than Pd, and of other nuclear transformations such as the transmutation of potassium to calcium (Dagani 1993), lending a certain justification to the occasional cries of "Alchemy!"
- Another well-known electrochemist, John Bockris, an early supporter of cold fusion, long well known to Fleischmann, accepted research funds to study nuclear transformations from a man who turned out to be a charlatan and who, in Bockris's laboratory, faked experiments purportedly producing gold (Begley and Levin 1994).
- A long-time supporter of Velikovsky's catastrophism (Jueneman 1989) claimed that cold fusion vindicated the scientific heresies about nuclear reactions promulgated by H. C. Dudley, a maverick professor of radiation physics at Illinois Medical Center (Chicago).
- There was a related suggestion that cold fusion supported the creationist belief that radiometric dating is inaccurate (Arndts 1989), and therefore the Earth might really be only about ten thousand years old.
- Cold fusion was made a central theme in the 1997 Hollywood movie *The Saint.*
- Articles about cold fusion appear in science-fiction magazines (Mallove 1997).

- The "Patterson Cell" was demonstrated at the American Nuclear Society meeting, Orlando, June 1997 (Mallove 1997). Electrolysis on metal-coated beads is supposed to produce nuclear transformations: increasing amounts of certain metals were detected after extended periods of time. This strikes electrochemists at first sight as a probable misinterpretation; extended electrolysis in the presence of otherwise undetectable amounts of impurities produces similar results (Zurer 1996).
- Claims were made of other unorthodox means of producing nuclear fusion: "capillary" fusion using pulses of current (Graneau 1994), or by cavitation in sonoluminescence (Pool 1994), a phenomenon itself little known and not understood (Glanz 1996b).
- It was proposed that Congress and the American Chemical Society should have strict guidelines for publishing "novel research no matter where it comes from" (Singal 1989), because articles rejected twenty years earlier are relevant to cold fusion.

Even support from apparently mainstream quarters could have less than welcome implications. No doubt Fleischmann and Pons were pleased to have their major paper on heat measurements published in *Physics Letters A* (Fleischmann and Pons 1993), but many people will have discounted its significance upon learning that the paper was published because the editor thought he could explain the results by his own pet but also unorthodox theory (Dagani 1993).

Fleischmann and Pons will not have been privately happy about the intellectual caliber of some who have flocked into "cold fusion." People who are attacked as they were find old professional friends turning a cold shoulder. When friendly support is then offered, those under attack may accept it before realizing that they might actually not have a great deal intellectually in common with those who offer the support.

Overreaction — Vicious and unfair attacks sometimes lead people to try to repair their reputations by taking legal action. Thus Pons, Fleischmann, and three Italian scientists sued *La Repubblica* because a book review "approvingly repeated Kohn's comparison of Pons and Fleischmann to 'fornicating priests,' and referred to cold fusion as 'scientific fraud'" (Holden 1996a).

The rejection of one's discovery is easily taken to indicate that there's something wrong with science as a whole. So the fight to justify a very specific piece of work may broaden into a full-fledged attempt at something better across the board, even tending toward megalomania: "frustrated by what Fleischmann calls 'consensus science,' they . . . 'are developing a new research organisation . . . to look at the science and engineering of the next century'" (Hodgkinson 1993).

Context

More than ten years after the announcement, it seems increasingly unlikely that there is a genuine energy-yielding process in "cold fusion." If there really isn't, then Fleischmann and Pons will have been very unlucky, especially perhaps in having Jones review their research proposal.

Still it's not clear that cold fusion is wrong or impossible. Competent theorists are able to sketch some lines along which such a thing could be explained, made to fit with *existing* physical theory: conceivably a means to tap into quantum fluctuations (Forward 1996), or some sort of concerted process analogous to laser action—science's ignorance of solid-state physics and chemistry is after all considerable.

Anyone who suggests that it is now easy to dismiss cold fusion as certainly erroneous (let alone that it ought to have been easy to do so at the beginning) should be required to read the remarkably even-handed and technically insightful book *A Dialogue on Chemically Induced Nuclear Effects: A Guide for the Perplexed about Cold Fusion,* published by the American Nuclear Society (Hoffman 1995); it details a veritable host of pitfalls that critics as well as proponents of cold fusion often fail to avoid. It is *odium scholasticum* to call cold fusion "pathological." Little if anything about it is very different from what happens all the time in science.

High-Temperature Superconductivity

High-temperature superconduction has not been called pseudoscience or pathological science; but it is instructive to compare its early years with those of N rays, polywater, and cold fusion.

Superconductivity is a kind of perpetual motion of electrons: an electric current flowing through a superconductor keeps going indefinitely, whereas ordinary electrical conductors offer resistance that steadily dissipates a portion of the electrical energy as heat. Superconductivity was discovered in 1911 in mercury cooled to about 4K; where K (for Kelvin; the degree sign ° is typically not used) is measured from the lowest possible temperature, the "absolute zero" temperature, 0K, where even the vibrations of atoms cease; 0K equals about –273°C or –460°F. Temperatures below about 80K are attainable only by great effort and experimental ingenuity, using the evaporation of exotic liquids like liquid helium to bring the temperature down. For widespread commercial applications, the "critical temperature" (T_C) below which superconduction occurs would need to be appreciably above 77K, which is the lowest temperature attainable with the cheap, readily available refrigerant, liquid nitrogen.

Over the years had come many false claims of high-T_C materials (Hazen 1988:33). In the mid-1970s, for example, magnetic anomalies suggestive of superconductivity at temperatures as high as 200K had made international headlines (Hazen 1988:8, 15). But by the mid-1980s, after some seventy-five years of effort, the highest *confirmed* T_C for superconductivity, in a metal alloy, was about 23K, reached in 1973. By then, however, "many scientists had abandoned research on superconductivity . . . assum[ing] that further substantial advances were unlikely, perhaps even physically impossible" (Hazen 1988:14, 30).

In the 1970s, though, the unknown unknown had yielded the surprise of superconduction in metal oxides, which are not usually thought of as even ordinary conductors of electricity; still it was only at very low T_C, below 20K.

And then in January 1986, Müller and Bednorz at IBM Zürich discovered superconductivity at a record high of 30K, moreover not in a metal but in a metallic oxide, "a material that by the conventional wisdom should not have been superconducting at all. At first, few physicists believed their results" (Hazen 1988:8).

Paul Chu at the University of Houston was the first to successfully repeat the work, making the same metallic oxide and confirming the T_C of 30K—but of fifteen samples he prepared, only two showed the superconductivity. Further work gave similar results: "out of each synthesis run only a few black chips became superconducting. . . . The great majority of samples showed nothing at all. Some samples would give clear signs of superconductivity, only to lose the effect after three or four days" (Hazen 1988:29). Varying the chemical composition, Chu attained a T_C of 73K—but the result could not be repeated on the same sample the following day.

Connectedness

The discoverers of high-temperature superconductivity were members of the scientific mainstream, but hardly central members. One doesn't normally expect such major scientific breakthroughs to come from industrial laboratories (a few places like the former Bell Labs notwithstanding; transistors and lasers were exceptional). At IBM Zürich, Müller's seniority allowed him to "work on just about anything he wanted," even so farfetched a quest which most experts viewed as misguided. His younger coworker "was forced into the uncomfortable situation of secretly synthesizing potential superconductors while simultaneously carrying out his authorized lab duties" (Hazen 1988:14–

15). The same dilemma, mutatis mutandis, hinders cold fusion and any number of anomalistic quests.

Paul Chu, who capitalized fastest on the initial announcement, was at a university with more ambition than achievement as a research institution—compare the Universities of Nancy (N rays) and Utah (cold fusion).

That superconductivity was found in metallic oxides may have made interdisciplinarity particularly vital. The traditional superconductivity community was based in physics; it had its own specialized theory and its own specialized methods for getting to very low temperatures. But the new developments called for different expertise: the chemical synthesis of metallic oxides, the analysis of samples that were very nonhomogeneous, the use of X-ray diffraction to determine structure. Work at high pressures may not have been crucial but it did play an important part. Paul Chu enlisted from various specialties people who were maverick enough to join this race to a farfetched goal (Hazen 1988). (Compare perhaps Robert Rines's enlisting of leading practitioners of sonar, strobe lighting, and photography and especially underwater photography in the search at Loch Ness, resulting in some of the most interesting data ever obtained there.)

When so many disparate specialties become centrally involved, it is not obvious to what extent the subject is connected or disconnected from the scientific "mainstream."

Although before the breakthrough at IBM Zürich the traditional superconductor community may have been skeptical, and even remained so for some of the hectic and publicity-saturated months thereafter, the validity of the discovery was soon generally accepted. The award of the Nobel Prize a year after the announcement was extraordinarily rapid recognition. Perhaps the facts were simply so compelling: after all, many groups in many places had been able to achieve even higher temperatures than Müller and Bednorz, with the same type of substance.

As to *theory,* high-T_C superconduction has no connection with the mainstream since there *is*—as of 1996 (Service 1996)—no satisfactory theory to explain it. Those who criticize parapsychology for having inexplicable results might ponder the fact that many competent scientists and many businesslike industrial organizations have devoted considerable effort over more than a decade to empirical work on substances for whose properties we have no satisfactory explanation. On the other hand, the *methods* used in high-T_C superconductivity studies are all quite well established in their respective specialties. The *facts* may be in dispute as each individual new claim is made, but by now the general claim is well substantiated, with ample instances of multiple replications of certain materials with T_C well above the commercially significant temperature of 77K.

Illustrative Points

Facts — With N rays, the dispute was over radiation, hence largely over measurement techniques. Without Wood's drastic demonstration by removing essential bits of the apparatus, proponents might have continued for a longer time believing Blondlot's claims even if many other groups couldn't repeat them. With cold fusion, the right conditions are sought for a process to be observed, for energy emission to occur, and one can suggest many possible shortcomings in the attempts of those who fail to achieve it. But with polywater and with high-T_C superconductors, at issue were properties of actually available bits of material, and resolution could come comparatively quickly as various talents and resources all joined in examining those tangible specimens—not always exactly the same as one another, but close enough because they exhibit the essential features.

Reproducibility — Chu's early observation of a T_C of 73K that could not be reproduced is reminiscent of the unique experience of Fleischmann and Pons in finding an apparatus to have exploded and the electrode having burned a hole in the floor. The inability to find superconductivity in each sample, though all were prepared *apparently* in the same way, is surely comparable to the lack of reproducibility of cold fusion.

The field of superconductivity has generated a number of claims that have still not been reproduced years later: "Unconfirmed reports of new compounds . . . abounded during the spring and summer of 1987. A Michigan company reported 155K. . . . A North Carolina research team claimed to have isolated a 240-K material. And dozens of researchers were tantalized by transient effects suggestive of superconducting temperatures up to—and even above—room temperature! But none of these observations was easily reproducible" (Hazen 1988:258).

The major difference is that *some* sort of reproducibility was attained in superconductors after weeks or months of effort, whereas with cold fusion it has been a matter of years without, yet, *any* satisfactory reproducibility. That's a matter of degree, surely, not of kind. Yet those who made these unconfirmed claims about superconductors have not suffered the calumny that has been heaped on cold-fusion claimants.

The Media — The fuss over high-T_C superconductors during the first months was very like that over cold fusion, including not only a lot of media attention and "science by press conference" (Felt and Nowotny 1992) but also at least one professional meeting so swamped by would-be listeners that many could not get in (the "Woodstock of Physics" at the annual meeting of the American Physical Society).

Ethics — Fleischmann and Pons have been criticized for not revealing full details of their procedures immediately. But their reticence is slight compared to the competitive maneuvering, attempts to mislead competitors, and dishonesty with journals that is part of the superconductor story. Scientific papers submitted to peer review contained deliberate errors to throw off possible competitors, the errors being corrected only at the last moment before actual publication (Felt and Nowotny 1992; Hazen 1988:part 1, chapter 4).

One of those mischievous errors even misled for a while a colleague of the man who perpetrated the falsehood (Hazen 1988:142). Coworkers went separate ways after disagreeing over speed versus sound science (Hazen 1988:58–59), and graduate students suffered thereby. Allies and colleagues mistrusted one another on the basis of rumor and misunderstanding (Hazen 1988:171, 208).

There was "a flood of hastily published articles, many of rather poor quality, and new journals that speeded up the publication process at the price of abandoning any referee system. Commercial (and expensive) 'newsletters' appeared, scientific results were announced first in the press to gain a few days on other groups, complete secrecy abounded in research even toward close colleagues. . . . [There was] fierce competition on national and supranational levels . . . as well as within individual groups" (Felt and Nowotny 1992:521). Whereas between 1983 and 1986, the *Science Citation Index* counted about 900 articles per year with "superconduct" in their title, in 1987 there were 2,450.

The sleazy, overcompetitive behavior of many of the participants has not, however, been criticized in the same manner as flaws in the behavior of Fleischmann and Pons have been. Why the difference? The only obvious reason is that the consensus accepts high-temperature superconductors as real but not cold fusion. Yet is unethical behavior any more acceptable when one is substantively right than when one is wrong?

The shenanigans in the superconductor race are all the more distasteful because only speed was at issue, not great originality or insight or even competence. Everyone was varying the elements and their composition in obvious ways. After the veil of secrecy lifted somewhat, it became clear that at least four independent and competing groups of workers had achieved much the same thing within a week or so of one another: "virtually every experiment and idea had been duplicated in these first hectic months of 1987" (Hazen

1988:241). More than fifty papers were published on the structure of the two forms of the "magic" material. But at the meeting at which everyone made their announcements, the moderator "announced the rules like a referee at a prize fight" (Hazen 1988:232) and one particularly self-serving statement "drew a chorus of boos and catcalls" (Hazen 1988:237).

In cold fusion, that the discovery came from other than one of the top research institutions lent a certain flavor to some of the attitudes displayed; with superconductors, that Paul Chu worked at non-elite Houston and those who prepared his material at non-elite Alabama lent a similarly unpleasant flavor to some of the attitudes displayed (Hazen 1988:223).

These unpleasantnesses, I suggest, have little if anything to do with whether it's science or pseudoscience that's going on. Unethical behavior associated with knowledge seeking stemmed primarily from the behavior of news-hungry media and greedy universities.

More Than Meets the Public Eye

A host of other claims in science, as major as cold fusion or polywater, have caused less furor but have nevertheless captured varying degrees of public attention and varying degrees of scorn from within the scientific community.

Whether or not a scientific dispute gains public attention depends on the media more than on the nature of the scientific issue at hand. Readers of *Science, Nature, New Scientist,* and the like, but not those of the popular media, knew of the claims made for gyroscopic devices that could counteract gravity (Holden 1990), an effect that could be as consequential as cold fusion or high-temperature superconductors. The existence of a fifth force (Franklin 1993)—in addition to the weak and strong nuclear forces, gravity, and electromagnetism—would have been as major as just about any scientific revolution, yet the controversy was little attended to by the press.

In these cases as with N rays and polywater, the issues were resolved in about the same length of time, about half a decade. Other puzzles may not be ripe for solution and can remain controversial for much longer, like Wegener's notion of drifting continents for nearly half a century.

As our knowledge in a given area deepens, it may become more and more difficult to distinguish genuine anomalies from meaningless glitches (Glanz 1996a); in high-energy or particle physics, for example, "anomalies are bound to proliferate . . . when accelerators have finished up their data-collecting runs. . . . 'Some people wait, analyze these carefully, and they disappear . . . and other people publish and say it may be a new phenomenon'" (Taubes 1996:474).

Here are just a few of the current controversial bits of science that the media have *not* chosen to make a major fuss about:

Prions

Little was heard by the general public about the controversy over "prions" until Stanley Prusiner was awarded the Nobel Prize in Medicine in 1997 (Rhodes 1997). In 1982 he had suggested that "*PR*oteinaceous *I*nfectious particles (*ONS*)" were the cause of "spongiform encephalopathies"—diseases in which the brain becomes riddled like a sponge with holes and with lumps of "junk protein."

Though the incidence of these diseases suggests that they are transmitted among people, there was no obvious infectious agent doing the transmitting. These diseases include

- kuru, which afflicts humans in Papua New Guinea. Cannibalism has been suggested as the transmitting mechanism.
- scrapie, a disease of sheep.
- bovine spongiform encephalopathy, B.S.E., "mad cow disease," which caused British beef to be banned from Europe during the 1990s. Both scrapie and B.S.E. are thought to be passed on when sheep and cattle are fed protein supplements manufactured from animal waste.
- Creutzfeldt-Jakob, a rare disease of humans characterized by progressive dementia and loss of muscle control. The observation in Britain of increased incidence of this syndrome led to speculation that it may have come through eating beef from cattle infected with B.S.E.

One of the mysteries is that the incubation periods for these illnesses appear to be extraordinarily long, a matter of years between the initial infection and the appearance of symptoms. By contrast, bacterial infections and virus infections produce symptoms in days or even hours.

In 1976 Carleton Gajdusek had received the Nobel Prize in Medicine for showing kuru to be infectious and owing to ritual cannibalism of relatives' brains. He thought the infectious agent must be a virus, but a new type that could remain dormant for years before becoming active: a lenti-virus or "slow virus." The mechanism or reason for dormancy and subsequent activation remained unknown.

Prusiner's concept that the infectious agents are proteins is based on experiments in which infectious material survives treatments that typically inactivate viruses. However, there is no theory to explain how proteins can cause other proteins to change their structure or shape.

Nobel Prizes are not usually awarded until the consensus in the scientific community is overwhelming that the discovery is valid. With prions it was otherwise: "only four of the fourteen major . . . research labs that . . . work on the infectious agent wholeheartedly espouse prion theory; [though] none . . . consider it unlikely, and one is undecided." The director of one of the labs made a public objection when the Nobel award was announced: similar clumps of abnormal protein as in the spongiform encephalopathies also occur in Alzheimer's disease, diabetes, and rheumatoid arthritis—why should prions uniquely be transmittable when these others are not (Rhodes 1997)?

HIV

AIDS (Acquired Immune Deficiency Syndrome) was first recognized as a breakdown of the immune system: people were dying in unusually large numbers from otherwise very rare diseases, and particularly within homosexual communities in San Francisco and New York. A gripping account of how the implications of these deaths were eventually recognized is Randy Shilts's *And the Band Played On* (1988).

Since the early 1980s, all public policy—notably including the allocation of research funds—has been based on the premise that the retrovirus HIV (Human Immunodeficiency Virus) is the single, necessary and sufficient cause of AIDS. Despite the great publicity given to AIDS over the last decade or two, the media have chosen to be virtually silent over the view held by a substantial body of competent scientists (Group for the Scientific Reappraisal of the HIV-AIDS Hypothesis 1995) that HIV is *not* the cause of AIDS.

The mainstream can be justly accused of doing in this instance what pseudoscientists are typically accused of: selecting from the data only what supports their own view; fudging statistics (for example, by changing the definition of AIDS more than once without appropriately adjusting the reported incidence and death rates); or evading relevant questions put to them (Duesberg 1996; Lauritsen 1993; Root-Bernstein 1993).

If HIV does cause AIDS, a continuing mystery is the very variable and often very long incubation period. The conventional wisdom calls HIV a Gajdusek-type "slow virus"; but if the type-specimen disease caused by a lentivirus, kuru, is actually caused by prions (see above), is then HIV the first and only lenti-virus?

Redshifts

The characteristic spectral lines of the elements observed outside the solar system are typically at longer wavelengths than on Earth—they are "redshifted" to longer wavelengths. According to the conventional wisdom, the degree of redshift is a measure of the rate at which the observed star is moving away from us and also a measure of its distance from us, under the current theory of a Big Bang and an expanding universe. Very little attention has been drawn by the media to the view espoused by a few highly accomplished astronomers, that these interpretations are not always valid. Halton Arp was ostracized by the community of American astronomers for pressing evidence that ought to be interpreted, he said, as showing calculations of distance based on redshifts to be grossly in error in certain interesting cases, quasars in particular (Arp 1987, 1998). Arp had to move to Germany because he was denied further use of American telescopes.

Quasars are acknowledged enigmas because the origin of their enormous energies is inexplicable under current theories. If Arp is right, then quasars could be much closer to us than now believed, which would go some way to resolving the energy dilemma—they wouldn't be emitting as much radiation as is now calculated. In recent years, several other people have uncovered further evidence that there is something wrong with current interpretations of redshifts; for example, redshifts apparently cluster around certain preferred values (Flam 1992a; Matthews 1996).

Origin of Life

The media have not publicized the controversies or dilemmas about the origin of life on Earth, which seems a little curious given that so much publicity surrounds controversies about evolution and creationism, which would seem to be inseparable from questions about the very beginning of life.

Most popular expositions say in rather vague terms that such molecular species as amino acids can be produced when ever-present things like water, carbon dioxide, and nitrogen are exposed to electric sparks or ultraviolet radiation, and that this is a sufficient indication that life originated on Earth from inanimate materials. Yet if truth be told, there is no consensus on the matter in the scientific community (Shapiro 1986). Moreover, some of the ideas bandied about are quite as fantastic as those of, say, Velikovsky—for instance, that life really got going chemically in outer space (Hoyle 1984; Hoyle and Wickramasinghe 1978), or that life on Earth was seeded from elsewhere possibly by spores (Crick 1981)—incidentally, a recurrence of notions extant a century ago. Recent speculation includes the intriguing idea that life flourished first deep within the Earth and only later at its surface (Gold 1999).

The lack of consensus and understanding in the scientific community on this matter seems to have stimulated a deceptive attempt to discredit scientific work on evolution and the origin of life. The Gene Emergence Project (2000) is preparing to offer "The Origin-of-Life Prize" of $1.35 million "for proposing a highly plausible *mechanism* for the spontaneous rise of genetic instructions in nature sufficient to give rise to life." But several details of the preliminary announcement mark this as less an honest offer than a Trojan horse intended to make plain that science has no such plausible mechanism to offer. The announcement put forth

- An emphasis that the genome carries "conceptual" information: "Which of the four known forces of physics, or what combination of these forces, produced conceptual information, and how? What is the empirical evidence for this kind of information spontaneously arising within Nature?"
- The assertion that teleology is a fact: "The simplest known genome's apparent anticipation and directing of future events toward biological ends."
- The argument that existing scientific theory is insufficient: "How did stellar energy, the four known forces of physics, and natural process produce initial information/recipe using direct or indirect code?"
- Reference to the creationists' latest assertion, that certain features of living systems display an "irreducible complexity": "Proposing a mechanism that explains the origin of life ought not consist of 'defining down' the meaning and essence of the observable phenomenon of 'life' to include 'nonlife' in order to make our theories 'work.'"
- The typical creationist misrepresentation of the Second Law of Thermodynamics: "Thermodynamic realities must be clearly addressed, including specific discussion of any pockets of momentary exception to the Second Law of increasing entropy and of the tendency for existing information to deteriorate in both closed and open systems."
- The assertion that the "Origin-of-Life Foundation should not be confused with 'creation science' groups," together with the absence of any names of individuals associated with the foundation: "The Board of Trustees consists largely of volunteer accountants, CEO's and owners of for-profit businesses who have a life-long interest in life-origins research."

Frontier Trash

The existence of popular trash in anomalistics, disseminated largely for commercial reasons, has parallels in the mainstream in the form of announcements of individual "research results" whose validity is highly suspect even at first sight and whose aim appears to be not furthering knowledge but gaining visibility, grants, or venture capital: "A gene involved in the search for new things" (Benjamin et al. 1996; Ebstein et al. 1996) and genes for excitability and impulsiveness (Anon. 1996b), alcoholism (Amato 1991), happiness (Holden 1996b), homosexuality (Pool 1993a), manic depression (Berrettini et al. 1995; Stine et al. 1995), schizophrenia (Straub et al. 1995), TV watching and divorce (Holden 1992). Low cholesterol may promote violence (Holden 1995). Stress causes hardening of the arteries (Anon. 1996a). "Stormy weather appears to trigger deadly bleeding strokes in men, often in the late fall, while women are more vulnerable in the springtime" (Anon. 1993). This is "junk science."

In many or all of these cases, it should be self-evident that at least two things are wrong: the studies didn't test for all the possibly confounding variables, and correlation is presumed to prove causation. The prevalence of "news coverage" of this sort shows what a vast, ready market there is for sensational news, especially if it has apparent relevance to human health or happiness. It is not obvious that indiscriminate reportage about ufology or parapsychology is more harmful or more deplorable than indiscriminate reportage that advises people to eat this, but the following week not this but that, or to relax immediately on pain of dying sooner. The warnings occasionally issued against taking seriously the latest health recommendations are also like the warnings coming from debunkers of pseudoscience, for example (Reese 1995): "a coalition of food and nutrition people issued . . . 'a list of 10 red flags to help consumers spot "junk" science.' A lot of dietary advice in the media is based on epidemiology . . . [which is] 'a crude and inexact science. Eighty percent of cases are almost all hypotheses. We tend to overstate findings, either because we want attention or more grant money.'"

The ten red flags were

1. Recommendations that promise a quick fix
2. Dire warnings of danger from a single product or regimen
3. Claims that sound too good to be true
4. Simplistic conclusions drawn from a complex study
5. Recommendations based on a single study
6. Dramatic statements that are refuted by reputable scientific organizations
7. Lists of "good" and "bad" foods
8. Recommendations made to help sell a product
9. Recommendations based on studies published without peer review
10. Recommendations from studies that ignore differences among individuals or groups

Most of this junk science about diets and genes is promulgated by accredited members of the scientific community through regular channels of scientific publication (subsequently picked up by the media, of course, who were assisted by press releases accompanying the scientific publications). What, if anything, distinguishes this stuff from commercial exploitation of trashy anomalistics?

"Pathological" Science

Debunkers don't hesitate to call something like ufology "pseudoscience." Matters like cold fusion or polywater that arise entirely within the scientific community, though sometimes also called pseudoscience, are perhaps more commonly labeled "pathological" science.

The expression comes from a talk given at the General Electric Research Laboratory in 1953 by one of the greatest of physical chemists (1932 Nobel Prize), Irving Langmuir. He addressed "the science of things that aren't so," using as examples the Davis-Barnes Effect, N rays, mitogenetic rays, the Allison Effect, extrasensory perception, and flying saucers.

Langmuir's remarks must have made a considerable impression, for fifteen years later a transcript was issued as a General Electric research report (Langmuir 1968)—an edited transcript not of Langmuir's original talk but of a recording made by Langmuir in March 1954. That Langmuir made this recording several months after the lecture also suggests that his talk had generated considerable interest. That 1968 research report was republished in 1985 in *Speculations in Science and Technology* (Langmuir 1985) and yet again in 1989 in *Physics Today* (Langmuir 1989).

Langmuir's lecture seems to have fitted the Zeitgeist of the 1950s. It was at about the same time that Martin Gardner published his classic survey of pseudoscience (Gardner 1957). The impetus that World War II gave to science and technology seems to have been accompanied by a surge of interest in fringe science—pseudoscience and real science may be antithetical in content but they seem to play to the same societal concerns (Bauer 1986–87).

Langmuir asserted six signs of pathological science:

1. The magnitude of the effect is substantially independent of the intensity of the causative agent.
2. The effect is of a magnitude that remains close to the limits of detectability; or, many measurements are necessary because of the very low statistical significance of the results.
3. It makes claims of great accuracy.
4. It puts forth fantastic theories contrary to experience.

5. Criticisms are met by ad hoc excuses.
6. The ratio of supporters to critics rises up to somewhere near 50 percent and then falls gradually to oblivion.

The 1985 version added to Langmuir's examples of pathological science "photomechanical and electromechanical effects . . . being reported with increasing frequency in papers from laboratories around the world" (93) though they conformed to the first five of Langmuir's characteristic symptoms. Sure enough, it turned out (Hall 1968:481; Hanneman and Jorgensen 1967) that those photomechanical and electromechanical effects were of "subjective nature." The epilogue in the 1985 version also suggests that Langmuir might well have included the canals of Mars and water dowsing (Foulkes 1971)

> as further well-known examples of phenomena that exhibit the characteristics of pathological science. The initial determination of the astronomical unit by radar echoes from Venus and its confirmation by an independent laboratory evidently suffered from the difficulties of dealing with observations near the threshold of detectability [here Langmuir cites Evans and Taylor 1959; Price et al. 1959; Thomson et al. 1961; subsequent bracketed citations in this extract repeat Langmuir's notes]. The subjects of polywater [Levi 1973], the effects of magnetic fields upon biological development [Kolin 1968], and reports of the detection of gravity waves [RJC 1973] also make interesting reading in this context.

Langmuir's talk was timely and made interesting suggestions. However, those stand up no better as diagnostics than do other suggested criteria for distinguishing science from pseudoscience (Bauer 1984a:145–46). The ratio of supporters to critics of extrasensory perception, for instance, has not fallen gradually from 50–50 to nil; fantastic theories contrary to experience could characterize quantum mechanics, black holes, the Big Bang, and much else in mainstream science; thinking up ad hoc modifications of theories to preserve them is now recognized as standard practice in mainstream science (Lakatos 1976).

Langmuir's disquisition and the fact that his diagnostics don't work are of some continuing significance because many practicing scientists, especially physicists, have come across one or another of the published versions and taken it as authoritative. So these diagnostics tend to be used as freely, superficially, and misguidedly as the notions that science has to be done by the scientific method and that scientific theories must be falsifiable. The *Science Citation Index* even in the early 1990s showed several references per year to Langmuir's seminar, and this is surely a proverbial iceberg-tip because so much of the discussion of anomalistics, pseudoscience, and pathological science is carried on in magazines that are not scanned for the *Science Citation Index.* For example, in *Zetetic Scholar* in 1980, Langmuir's "pathological science" was cited even by that experienced and knowledgeable critic of parapsychology, Ray Hyman (Hyman 1980).

One illustration of the unfortunate influence of Langmuir's remarks is the article in *American Scientist* (Rousseau 1992) that discussed polywater, cold fusion, and "infinite dilution" studies of the effectiveness of certain biological agents as "case studies of pathological science." These lapses were ascribed to "loss of objectivity," as though scientists are not being pathological only so long as they do practice objectivity—an absurdity, given that human beings are incapable of being reliably objective. Objectivity is spawned not by individuals but by the mutual critiquing that does decrease subjectivity (Bauer 1992a). Moreover the author of that article (Denis Rousseau) may himself have displayed less than objectivity. He had played a role in the polywater affair, maintaining that polywater was ordinary water contaminated with, among other things, human sweat; but that theory is not generally accepted (Franks 1981:135).

It's quite usual for such writings as Langmuir's, about "pseudoscience" or "pathological science," to remain without criticism for decades. Few have an interest in defending those who were maligned or in pointing out how inaccurate a picture of scientific activity is thus allowed to stand. Long past due is a reconsideration of what Langmuir set in motion, the lumping together of error in and outside science on the basis of superficially plausible yardsticks that turn out in practice to be unusable for making the necessary distinctions. For instance, Richard Rhodes has pointed out that Langmuir's measures of pathological science fit nicely the discovery of prions for which Prusiner received a Nobel Prize (Rhodes 1997).

Is there really justification for calling N rays the same sort of thing as flying saucers? If so, what is that justification?

5 Mainstream or Unorthodox? Straddling the Pale

Some subjects seem to have a foot in both the scientific and the anomalist communities. A single subject that is emblematic of this, and thereby emblematic of some anomalistic topics too, is the relationship between electromagnetic energy and life. It has been at times entirely mainstream, at other times beyond the pale, quite often both at the same time; and it still occupies a fascinating borderland, some aspects accepted in the mainstream while others remain heretical.

The relation between electromagnetism and life is not a purely "scientific" issue; it connects inextricably to medicine and thus to social science. There are other topics too—archaeoastronomy, for example—that partake of both natural and social science and where the distinction between orthodox and unorthodox is far from clear.

Bioelectromagnetics and Life

A major point of controversy has long been between "holistic" bioelectromagnetism—asserting that electromagnetic energies are effective at the level of the whole organism—and the contrary belief that there are only specific individual bioelectromagnetic processes at the chemical or cellular level. In earlier days, the dichotomy was exemplified by the use of electric baths for whole-body "galvanization" (charging) as opposed to stimulation of specific muscles with locally applied electrodes. Nowadays the opposition is between notions of a "whole-body" electric field or pattern of voltage that influences among other things the regeneration of limbs, and purely local electric influences, say, across cell membranes.

I have not found any comprehensive work that traces how "electromagnetism and life" has moved in and out of mainstream accreditation over the centuries, but hints can be found in a number of histories. A magnificent resource for studies in this field is the Bakken Library and Museum in Minneapolis. Many of the following historical snippets are from *Electricity and Medicine—History of Their Interaction* (Rowbottom and Susskind 1984; page numbers given in parentheses in this section without another source citation refer to that work). An earlier but still useful review of the history of electrotherapy is by Licht (1959).

> The functioning of all living organisms, from the simplest cell to the human body, depends on electricity. Attempts to understand how electricity and life interact date back to antiquity . . . [and] physics often lagged behind medicine in advancing this understanding. Electric fishes were known and used before the source of their strange power had been discovered. Animal muscles were among the earliest indicators of electric phenomena, and the intensity of the electric shock perceived by an investigator served as a rough calibration of the quantity of electric charge. . . . Wound currents were detected long before they could be reliably measured. . . .
>
> Practical remedies adopted with next to no scientific rationale, like 19th-century electrotherapy, were wide open to abuse by charlatans, to the point of almost totally discrediting such methods. . . . [T]he medical uses of electricity—and magnetism—have attracted more than their share of quacks over the centuries. (260–61)

When inanimate objects were thought capable of harboring spirits or souls, Thales (about 600 BC) had suggested that magnets had souls because of their power to move iron (1). The production of shocks from electric ray-fish had "therapeutic use in gout, persistent headaches, and so on" (35). During the sixteenth to eighteenth centuries, many observations were made of the ability of human bodies to be electrified and to conduct electricity.

Early in the eighteenth century came speculation that the "animal spirits" or "nervous fluid" connecting the brain with the muscles might be electrical in nature (32). Late in that century, Galvani was using electrical apparatus for experiments with muscle-nerve preparations from frogs (35). In the middle of the eighteenth century, the connection between electricity and life was confirmed by such observations as the killing of sparrows by electric discharge from a Leyden jar, one of the common devices used for accumulating electricity. The ability to stimulate specific muscles was discovered, and there were successes as well as failures in attempts to treat paralyzed limbs (15–16).

Johann Krüger, professor of philosophy and medicine, in 1743 suggested that health could be maintained or restored by the passage of electricity. An increased pulse rate was observed upon electrification, and finger stiffness was cured by electric treatment (7). Electric cures of epilepsy, deafness, swollen glands, headaches, and rheumatism were reported, but often these could not be repeated. "By the 1770s and 1780s medical electricity had achieved a certain measure of official recognition in Britain, the technologically most advanced country, as a method to be tried for cases that did not respond to other therapeutic procedures" (23); indeed electricity was used even to revive people apparently dead, for example from drowning. An analgesic effect of static electricity was "well attested" (29–30).

It was also reported that electric treatment could make drugs effective even if they were present only in a sealed container in the electrical machine; but others were unable to repeat this observation (18). A recurrence of this in the 1990s is Benveniste's claim to be able to transmit by electromagnetic radiation the efficacy of homeopathic dilutions of immunological agents (Aïssa et al. 1997; Sheaffer 1998).

The mixture of acceptance and rejection, of mainstream and heresy, is illustrated by the fact that toward the end of the eighteenth century, special institutions for electrical treatment were established in Britain. At the same time, "the notorious quack, James Graham," introduced a Temple of Health in which brilliant glass cylinders, brass rods, and a "tremendous pyramid of conductors" produced "an astonishing fiery dragon" and "the electrical fire" (27–28). Infertile couples paid Graham exorbitant sums to make love on the "electro-magnetic bed" in his "celestial chamber" to the strains of an orchestra playing outside, while a pressure-cylinder pumped "magnetic fire" into the room (O'Conner 2000). T. Gale, an American physician, practiced "medical electricity" for a couple of decades, apparently sincere in his beliefs though his writings on "animal, vegetable, and astronomical electricity border on fantasy" (29); nevertheless, his report to have cured a few people of madness by administering strong shocks anticipated the electroshock treatment (re)introduced in the 1930s.

A consequential controversy about the nature of the connection between electricity and life was between Galvani (1737–98) and Volta (1745–1827): did nerves and muscles generate electricity, or did they merely respond to external applications of it?

For some time, the latter view (associated with Volta) was dominant and colored the interpretation of observed electrical interactions with biological specimens; the theory of "animal electricity" was forgotten for about the first half of the nineteenth century (44–45). It remained active at least in the popular imagination, however, as illustrated by Mary Shelley's *Frankenstein,* published in 1818 (Glut 1973:12–15; Small 1973:35, 45). And applications continued: a test of death was whether muscles would contract when electrically stimulated, and electrical treatment was used for resuscitation as well as to treat paralysis. Acupuncture had been introduced in France around 1820 and soon was combined with electrical stimulation using acupuncture needles (54).

Following discoveries in physics and chemistry during the early parts of the nineteenth century that connected electricity and magnetism, magnetoelectric medical devices were used to treat nervous diseases. These machines were being produced even in the twentieth century (59). However, at the same time such "electrotherapy had fallen into general disfavor among legitimate medical practitioners. Except for a very limited number of enthusiasts, it tended to be used only as a last resort. One reason was that toward the end of the eighteenth century a number of irrational forms of treatment had become popular, which superficially appeared to be related to electrotherapy," notably the "animal magnetism" of Franz Mesmer and the "magnetic tractors" invented and sold by Elisha Perkins in America (63).

> Certain forms of magnetic healing continued to be widely practiced during the first half of the 19th century, including methods employing actual (permanent) magnets . . . Some of the practitioners were pure charlatans; others were sincere. . . .
>
> At this period galvanism also provided a fertile source for exploitation by quacks. Belief in the mystical power of metals had lingered on from ancient times, and the discovery of galvanism gave such ideas a new impetus. Large numbers of amulets incorporating zinc and copper were manufactured . . . [as well as] headbands, belts, and so on, formed of alternate disks of copper and zinc attached to some kind of fabric. . . . 'Perpetual Electromotors and Galvanic Supporters' and 'Electro-motor Teething Necklace for Children' are examples of a form of quackery that continued to find a ready market from the 1840s until the early years of the 20th century and beyond; even today some educated people wear copper bracelets for 'therapeutic' reasons. (64)

To the present time there persists greater antagonism to the notion that magnetic fields can influence living systems than that electric or electromagnetic can do so. Trying to ascertain the facts of the matter is greatly complicated by the circumstance that the application of magnetic fields almost invariably has an electrical as well as a magnetic effect whenever the magnetic field is varying in strength, or when there is relative movement between the magnetic source and the conducting object being magnetized. So it's generally possible to suggest that any effect observed under magnetization is actually owing to electrical currents induced by the magnetic field.

"In the late 1830s and early 1840s electrotherapy gradually returned to greater favor among medical practitioners, largely owing to the use of induced currents obtained from the magneto-electric and electro-magnetic machines" (66). Some hospitals had departments of electrotherapy (103). There were claims that the absorption of drugs could be enhanced galvanically, and also that harmful metals could be extracted from the body galvanically; and there were counterclaims and protestations at incompetent or unqualified practitioners (182). It was still a borderland that stretched toward science fiction; for instance, one surgeon and Fellow of the Royal Society went so far as to propound a vitalist-like doctrine of "the voltaic mechanism of man" (69).

The introduction of electrocardiology and electroencephalography confirmed the significance of the electrical properties of living systems. At the same time, radio transmission was being developed, and naturally there was speculation that "brain waves" might allow for psychic phenomena (165–67). Cardiology led to development of the defibrillator that is now in standard use although "as late as 1950 many physicians opposed this technique" (248). The development of the pacemaker also met with resistance like that of the American Medical Association in 1930: "reports of resuscitation by cardiac injections belong with the miracles" (249).

Since visible light is a form of electromagnetic radiation, it was natural that biological effects of visible light too would be studied. In Scandinavia, arc lamps were used to treat *lupus vulgaris* and other conditions, and a Nobel Prize was awarded for this work in 1903. But extravagant claims were also being made for light treatment (Reader's Digest Association 1981:250–51); and that continues in the New Age—"Color therapy is a form of photo-therapy using color to influence health and to treat various physical or mental disorders," according to 1 of 879 items returned from the Internet by Alta Vista searching under "color therapy" in 1997. Analogous to or derivative of color therapy is gem therapy: rubies give out "warm rays" whereas emeralds emit only cold ones (Anon. 1997a).

In the twentieth century, improvements in public hygiene, the invention of vaccines, and the introduction of sulfa drugs and antibiotics made electrotherapy redundant for a number of conditions and in general relegated it to obscurity until the 1960s (178–79). "Abuse of the method, in treating a whole range of human ills from cancer to colds toward the end of the nineteenth century, caused a backlash as science began its march to become the backbone of medical practice. Soon electrical stimulation was looked upon as quackery" (Bassett 1995:262). "A major evaluation of American medicine . . . published in 1910 . . . [the influential Flexner Report] denounced the clinical use of electric shocks and currents, which had been applied, often over-enthusiastically, to many diseases since the mid-1700s" (Becker and Selden 1985:82). "Medical techniques have come to be tested as much against current concepts in biochemistry as against their empirical results. Techniques that don't fit such chemical concepts—even if they seem to work—have been abandoned as pseudoscientific or downright fraudulent" (Becker and Selden 1985:18). Loewi's discovery (1936 Nobel Prize) that acetylcholine mediates the transmission of nerve impulses across synapses had "resulted in the collapse of the last vestige of electrical vitalism" (Becker and Selden 1985:67).

Since the 1960s, however, bioelectromagnetics has come back into better repute in at least some quarters. Successful bone healing by electrical or electromagnetic stimulation has brought further research by electrochemists, biophysicists, and others as well as the orthopedists who were its pioneers; by 1976, "fourteen research groups had used bone stimulators on some seven hundred patients, for spinal fusions and fresh fractures as well as nonunions, all with seemingly good results" (Becker and Selden 1985:168).

Magnetic stimulation of the brain, tried by d'Arsonval in 1896 and by Thompson in 1910, was indubitably achieved in the 1980s at the University of Sheffield (Ueno 1994:vii, 1, 29). A few years later, magnetic stimulation of the heart was demonstrated (Nyenhuis et al. 1994). But still controversy and questioned unorthodoxy continued with, for example, claims that epilepsy could be cured by beaming into the patient's brain a magnetic field of the same intensity and frequency as that detected by magnetoencephalograms from the focus of the epileptic seizure (Crease 1989).

That studies of this sort are now somewhere in the scientific mainstream is attested by monographs from the publisher Birkhauser in its *Bioelectrochemistry* series; works by Baker (1989), Blank (1995), and Malmivuo and Plonsey (1995); and conference proceedings (Milazzo and Blank 1983; Ueno 1994) and journals (such as *Bioelectrochemistry and Bioenergetics* from Elsevier Sequoia) published or sponsored by mainstream organizations. Bioelectrochemistry is now among the topics of the highly respected Gordon Research Conferences. The 1986 presentations included "Genetic and Teratological Effects of EM Fields," "Bioeffects of Geomagnetic Fields," "Oncological Uses of EM Fields," "Bone Healing by EM Fields," "Wound Healing by EM Fields" (Anon. 1986); in 1996, "Biophysical Mechanisms of Magnetic Field Reception" and "Biophysical Mechanisms of Electric Field Reception."

On the other hand, that the field is still very much in flux is illustrated by the remark that bioelectromagnetism should not be confused with "bioelectromagnetics," which is "mostly concerned with biological effects of externally applied fields" (Foster 1996). Overlap with anomalistics is illustrated by the Bio-Electro-Magnetics Institute, whose affiliations include holistic and homeopathic medicine, the Foundation for the Study of Cycles, Bioelectronics Laboratories, the Center for the Study of Consciousness, the Light Research Foundation, and the Psychic Research Laboratory. The institute's newsletter features such pieces as "Studying the Electromagnetic Correlates of Laying-on-of-Hands Healing (Therapeutic Touch)," "Clinical Cosmobiology," and

"A Bioelectromagnetic Explanation (and Possible Solution) to Tinnitus" (Bio-Electro-Magnetics Institute 1989).

Electromagnetic senses are being discovered in a variety of species, for example, electroreceptors in the bill of the platypus. Reese writes that it's "probable that the animals rely almost entirely on electrodetection to find prey and assess obstacles on the bottom and changes in the depth of streams" (Reese 1987). The detection of prey may work by the sensing of the electrical activity associated with muscle activity in the prey (Reese 1987).

For more than a decade there's been a continuing controversy over possible harmful effects of low-level electromagnetic fields: is the risk of contracting cancer increased by use of appliances like electric blankets or hair dryers, or by being in the vicinity of power lines? "In the 1970s, it seemed absurd that EMFs . . . could be hazardous, even in the tiniest degree. Now it's a legitimate open question" (Pool 1990:1096). "Evidence suggests that cell processes can be influenced by weak electromagnetic fields" (Goodman et al. 1995:279) although "there have been some unsuccessful attempts to reproduce prior positive effects" and "there is no widely accepted biophysical explanation for the way weak EMFs produce noticeable changes in biological systems"; "the phenomena . . . do not have a simple dose-response relationship. Essentially none shows a linear increase with exposure length or intensity; many seem to saturate quickly once a threshold has been passed. . . . 'windows' of frequency or intensity have been observed." There are even data suggesting that weak magnetic fields alone can produce a biological effect (Goodman et al. 1995:280).

Such open technical issues that impinge on public policy bring controversy in which the technical issues overlap with clashing political ideologies, and both sides attract unwelcome extremist bedfellows; the possible danger of EMFs has seen the publication of books with titles like *Currents of Death* (Brodeur 1989).

The effect of electricity on the growth of plants had been noted in the late eighteenth century, albeit dismissed by others as fanciful (25); but in the 1990s, patents were filed by Life Resonance, Inc., for combinations of electromagnetic fields to increase the stress-resistance of plants (Andrews 1992). In both instances, entirely mainstream practitioners made the claims: Pierre Berthelon in the eighteenth century; in the twentieth, Abraham Liboff (physics, Oakland University), Bruce R. McLeod (electrical engineering, Montana State University), and Stephen D. Smith (anatomy, University of Kentucky). In between, however, there had also been claims about electrical signals from plants that are generally dismissed as spurious. Cleve Backster asserted that his polygraph could detect the "emotions" plants felt under various stimuli (Reader's Digest Association 1981:250–51). Scientists at the University of Colorado claimed that growing algal cells generated electrical fields that controlled the subsequent shape of the organism (Anon. 1980).

The Body Electric

Robert Becker exemplifies both modern scientific knowledge in bioelectromagnetics and its overlap with holistic views associated with alternative medicine. He tells the story of his work in a book (Becker and Selden 1985) whose title is also the heading of this section; page numbers cited below refer to that work except where otherwise indicated.

Among Becker's salient assertions and speculations is that the regeneration of limbs and organs will become possible once a better understanding of the body's electrical activity has been achieved. "I'm sure we can get the human heart to mend itself. . . . [T]here's hope that we'll soon be able to coax nerve cells into reestablishing the proper connections across the damaged section [of the spinal cord]. . . . [I]t seems likely that a similar stratagem could induce brain regeneration in animals normally lacking this ability" (202, 203, 213).

This is the sort of claim that one might be tempted to dismiss as science fiction or pseudoscience—were it not for the fact that Becker and others have in fact already achieved partial regeneration in such species as frogs and rabbits that don't normally regrow lost limbs (Becker and Selden 1985:19; Pilla 1974). Crucial to regeneration of limbs is the regeneration of nerves; the "notion that an imposed electrical field can enhance or direct nerve growth began with the work of Sven Ingvar in 1920" (Borgens et al. 1981). Growth in general may be influenced: "strong electric fields tend to accelerate the growth of male chicks for a few weeks after birth" (Reese 1976). Surgeons have known for at least twenty years that in children no more than eleven years old, if a finger is sheared off above the last joint, the fingertip "will *invariably* regrow perfectly in about three months" (156).

Another of Becker's ideas is that the movement of ions that transmits nerve impulses is only one small part of the body's electrical activity. Direct current (DC) or direct-voltage differences are transmitted or maintained by paths along the perineural sheaths that surround nerve fibers, which are not conductors of electricity so much as semiconductors. "The cells that biologists had considered merely insulation turned out to be the real wires" (237).

Here Becker assuredly parts from the mainstream. Especially when he continues that

- This DC network controls healing and regeneration. It's also "part of what many people mean by the soul" (182).
- The DC network also coincides with the meridians used in acupunc-

ture. Acupuncture points along these meridians are also nodes of voltage. The acupuncture meridians are electrical conductors that carry an injury message to the brain; if the message is blocked, no pain is felt—that's how the analgesic effect of acupuncture works (234).

- "The earth's surface and the ionosphere form an electrodynamic resonating cavity that produces micropulsations in the magnetic field at extremely low frequencies . . . [chiefly] at about 10 hertz (cycles per second)." Even very weak fields at this frequency of 10 Hz can dramatically restore such cycles as circadian rhythms. This is also "the dominant (alpha) frequency of the EEG in all animals" (247, 249).

  The origin of life may have been electromagnetic action on inorganic semiconductors, at a time when the distance of the ionosphere from the earth's surface tuned that "resonating cavity" to 10 Hz. The fact that living things use molecules of only one of the two possible mirror-image forms stems from the asymmetry of the Earth's magnetic field at the time. Mass extinctions at the end of the Permian and Cretaceous eras have to do with "frequent magnetic pole reversals . . . after a long quiescence" (259, 262).

  Again one is tempted to regard this as just too glib, altogether too many disparate matters tied too neatly together when the reality has to be vastly more complicated, just as dianetics is an oversimplified science-fiction sort of psychology. Yet it seems that most, perhaps even all, living things do indeed have the capacity to sense magnetic fields.
- "We don't know how electrically produced silver ions differ from ordinary dissolved ions, only that they do" (175)—a difference unknown to mainstream electrochemistry, which generates ions electrically all the time; this is uncomfortably reminiscent of claims about polywater (water formed in tiny capillaries is different from ordinary water), or Benveniste's claim to inject immunological activity into water by electromagnetic means (Aïssa et al. 1997).
- "Molecular dynamics, the simple gradients of diffusion, can't explain anatomy" (181). The slide from "not yet explained" to "unexplainable" is one of the trademarks of spurious science.

In form as well as substance *The Body Electric* shows a number of points of similarity with works of extreme unorthodoxy. Thus, no sources or references are given to enable readers to pursue or check mentions of people and publications. Moreover, obscure or even more unorthodox work is mentioned as supporting Becker's claims: supposedly proven instances of psychic healing, dowsing, extrasensory perception (27, 255, 264); Russian work on "variation of the bioelectric potentials in the regeneration process of plants" (61); the "field of life" or "L-field" of Harold Saxton Burr (whose work on this during the 1920s was self-published because the orthodox professional journals rejected it); the work of Frank A. Brown, who suggested that "rhythms in the earth's magnetic field served as timers for the rhythms of life" (107, 246, 248); the ideas of Francis Ivanhoe, who "found bursts of brain-size evolution at about 380,000 to 340,000 years . . . and . . . 55,000 to 30,000 years ago," correlated with magnetic-field changes (263–64).

Yet no sooner is one tempted in exasperation to stop reading—as extrasensory perception is suggested to result perhaps from some new sort of biomagnetic field and one again catches the whiff of pseudoscience—than Becker recounts having experimented with Kirlian photography and rejected it on the basis of the evidence. Becker may be willing to speculate far into the unknown, and he may be unwaveringly certain that some sort of holistic view has to be the right one, but he is also entirely committed to following where the facts seem to lead.

The book begins within the conventional wisdom but then extrapolates limited and specific new claims into a whole counterconventional system that hints of megalomania as it explains everything, including the most fundamental human questions. Becker writes:

> My research began with experiments on regeneration, the ability of some animals, notably the salamander, to grow perfect replacements for parts of the body that have been destroyed. These studies . . . led to the discovery of a hitherto unknown aspect of animal life—the existence of electrical currents in parts of the nervous system. This breakthrough in turn led to a better understanding of bone fracture healing, new possibilities for cancer research, and the hope of human regeneration—even of the heart and spinal cord—in the not too distant future. . . . Finally, a knowledge of life's electrical dimension has yielded fundamental insights . . . into pain, healing, growth, consciousness, the nature of life itself, and the dangers of our electromagnetic technology. (21)

Becker even claims, "We're dealing here with the most important scientific discovery *ever*—the nature of life" (327).

Becker's book conveys the belief that almost everything about Establishment science is wrongheaded and fossilized whereas almost all unorthodox ventures are on the right track:

—"The experiment itself was as simple as could be. The tricky part was getting permission to do it" (68).

—"Nevertheless, some researchers kept coming up with observations that didn't fit the prevailing view. Although their work was mostly consigned to the fringes of the scientific community, by the late 1950s they'd accumulated quite a bit of evidence" (82).

—In magnetic fields "we'd discovered the best possible anesthetic . . . but reactions by living things to magnetic fields were absolutely out of the question in America at that time" (114).

—"[I]t generally becomes obvious [sooner or later] that the iconoclast was right." Unfortunately this statement follows mention of a couple of iconoclasts who were *not* right, namely, Velikovsky and Reich.

Becker's remarks often take on an alarmist tone. The dangers of artificially generated electromagnetic fields are described in apocalyptic fashion, for example, a higher suicide rate near power lines; "electropollution is presenting us, and perhaps all animals, with a double challenge: weaker immune systems and stronger diseases . . . All life pulsates in time to the earth, and our artificial fields cause abnormal reactions in all organisms" (294, 328). In another passage he asserts, "It now seems feasible to induce catastrophic climate change over a target country" (326). Becker's public testimony about such matters brought retribution from various government and Establishment groups, including loss of research funds and facilities (chapter 15).

As so often, we are dealing with matters of degree. In the tone with which it draws attention to the conservatism of the mainstream, *The Body Electric* is intermediate between rank pseudoscience and mainstream science. There is more than just a grain of truth to the extreme conservatism of science; and orthodoxy *is* sometimes enforced by unethical tactics. Becker's account is not unbalanced; at several places he goes out of his way to praise individuals who are part of the Establishment but who fully met the classic standard of the objective yet open-minded scientist. I see no reason not to accept as plain fact and close to verbatim Becker's report of what he was told in a meeting circa 1960: "We have a very grave basic concern over your proposal. This notion that electricity has anything to do with living things was totally discredited some time ago. It has absolutely no validity, and the new scientific evidence you're citing is worthless. The whole idea was based on its appeal to quacks and the gullible public. I will not stand idly by and see this medical school associated with such a charlatanistic, unscientific project" (70).

Becker's "Postscript: Political Science" (330–47) contains much that cannot be gainsaid: that greed and prestige-seeking are—or have become—about as common among scientists as outside science; that there are all too many instances in academe and in research establishments of petty jealousy and dishonesty; that the influx of huge research funds has corroded academic values; that "If it's trivial, you can probably study it. If it's important, you probably can't" (333).

In considering the high degree of resistance Becker encountered throughout his career, it's worth bearing in mind that he was an M.D. working in the bureaucracy of the Veterans Administration and seeking funds from other government agencies as well. Researchers in universities and independent research institutes are often less hampered, as—some of the time—are some of those in some industrial research centers (for instance, leading to high-T_C semiconductors). Thus Robin Baker at the University of Manchester in Britain:

> From the beginning, for some inexplicable reason, the idea that any animal might be sensitive to the Earth's magnetic field has generated intense and heated opposition from the more conservative sections of the scientific community. . . . [N]ineteenth-century hypotheses concerning the sense of direction of humans and other animals retreated from prominence. It was 1965 before Merkel and Wiltschko . . . produced the first effective demonstration that . . . the European Robin . . . had a magnetic compass sense. Their results, after a brief series of skirmishes with opponents, eventually triggered similar demonstrations for more and more animals by more and more authors. (Baker 1989:x)

Baker's heresy has been to study *human* sensing of magnetic fields, which attracts opposition for no obvious reason: "Every animal seriously tested has been found to have a magnetic sense, and it now seems that the final search will be for an animal that is magnetically blind. It would be more surprising to discover that Man just happened to be that animal than to discover he was not" (Baker 1989:ix).

Baker cites initially dismissive comments from his peers: "He's just careless"; "To get Baker's results, you'd really have to be a sloppy scientist." Amusing is Baker's demonstration that some of the competitors who obtained results supporting his thesis nevertheless chose titles for their articles that masked the fact, such as "Attempts to Replicate Baker's Displacement Experiment," "Critique and Alternative Hypotheses," and "An Elusive Phenomenon." As Baker observes, the "atmosphere surrounding the study of human magnetoreception has, both publicly and less publicly, been just a little unpleasant" (Baker 1989:xi).

Nevertheless, despite such strong resistance to the work on animal navigation by magnetic sensing that he pursued from early in his career, Baker has attained the rank of Reader (corresponding in the United States to Research Professor). That he was less hampered and less harassed than Becker may be not only the difference between academic and government research but also a difference between Britain and the United States, Britain having a long tradition of respecting and even fostering intellectual diversity and eccentrics. Moreover, until Thatcherism, academics at British universities could count on some useful level of research support as a routine part of their academic employment; not needing to apply for specific grants, they could work on whatever seemed to them worthwhile without beforehand having to convince others of its merit.

Connectedness

Bioelectromagnetics has been a concern of a number of different disciplines, at various times and in various places. No single overall measure of its connection to the relevant mainstream communities could then be valid; rather, the historical sketch given above shows that a host of detailed historical studies wait to be made, to elucidate exactly how those who practiced or investigated various aspects of bioelectromagnetics at various places and times related to mainstream practitioners. There's a certain similarity between these checkered, changing relations with the mainstream and the contemporary circumstances regarding cold fusion.

National Variations — Bioelectromagnetism in medicine has run significantly different courses in different countries. Electrotherapy in the late nineteenth century was largely by galvanic current in Germany but by faradic in France (Rowbottom and Susskind 1984:chapter 7). In Britain electrotherapy was later getting back into favor; in the 1890s, the leading work was done in France (Rowbottom and Susskind 1984:181), including the centrally important observation that different frequencies of field had different effects: for example, frequencies of 100,000 to 10,000,000 could be safely applied while the same current at 100–150 cycles could be lethal (Rowbottom and Susskind 1984:128–29).

Sleep induced by electrical currents (electronarcosis) is widely used by legitimate therapists in France and Russia (Becker and Selden 1985:86) but is heretical in the United States.

Disciplinary Varieties — The mainstream is also not easily defined in disciplinary terms. One of the pioneers of electromagnetic stimulation of bone healing believes that progress will depend on cooperation among "physicists, engineers, biologists, geneticists, physiologists and physicians, but particularly chemists" (Bassett 1995:273); but there is as yet no agreement even over what is encompassed by "bioelectromagnetism." Thus in the Advances in Chemistry series, the 1995 volume, *Electromagnetic Fields—Biological Interactions and Mechanisms* (Blank 1995), has articles by thirty-four authors and coauthors; but none of those is even referred to in the 1995 text *Bioelectromagnetism: Principles and Applications of Bioelectric and Biomagnetic Fields* (Malmivuo and Plonsey 1995). The latter is written by authors (one in Europe, the other in the United States) whose discipline is described as "biomedical engineering"; in the former, contributions come from disciplines also described as biomedical engineering but, as well, bioelectromagnetics, biomedical science, health science, medical radiation research, and a variety of other medical, engineering, and science specialties (anatomy, cellular biophysics, epidemiology, neurobiology, orthopedic surgery, pathology, physiology; chemical, electrical, and computer engineering; biology, biochemistry, chemistry, geology, molecular biology, and physics).

Reproducibility: Connecting the Facts — Connectedness judged in terms of facts, methods, and theories again does not yield any simple, overall answer. As to the facts, the lack of reproducibility is salient; as with cold fusion, there have been a variety of claims either not reproducible or that no one has tried to reproduce.

Until a phenomenon is rather fully understood, it cannot be known what might interfere with the ability to repeat it successfully. The effectiveness of electromagnetic healing seems to have much to do with the particular waveform as well as its duration and magnitude (Markov 1994): for example, one can influence separately the rejoining of fractured bone, the local degeneration (osteonecrosis) of bone, and osteoarthritis, according to what electromagnetic impulse is applied (Bassett 1995). Tests of electrotherapy before such specificities were known would be frustratingly irreproducible, delivering some random mixture of positive and negative results. Various durations and magnitudes of signals would sometimes, by chance, entrain some of the efficacious frequencies or wave-form shapes but at other times not. Lack of reproducibility could make it seem that a will-o'-the-wisp was being pursued rather than a genuine phenomenon. The debunkers would have a field day.

Again, it now seems that cells can detect external EMFs whose magnitude is small compared to the random electromagnetic fields naturally present inside living systems. Apparently some coherence or resonance characteristics are involved that enable detection of the external field. At the same time, external random electromagnetic fields (EMF "noise") would swamp other external signals. Once again, experiments carried out without knowledge of these effects—effects that one would not normally expect—would be inexplicably, frustratingly irreproducible (Malmivuo and Plonsey 1995). The disbelievers could rhetorically ask in sneering fashion, how anyone could expect such a small external signal to be effective in the presence of so much internal random noise.

Even seemingly reasonable attempts to make results more reproducible can have the opposite effect if the attempts are based on erroneous speculation about the phenomenon—erroneous speculation that may be based on the conventional wisdom. For example, in electromagnetic experimentation in physics and chemistry, great care is taken to achieve uniform fields; but in exploratory work in *bio*electromagnetics, "attempts at uniformity may in themselves reduce the probability of observing an EMF effect if a bioresponse depends on nonuniformities, transients, or interactions between more than one type of field. . . . Some experimental evidence indicates that the greatest effects may occur with onset or removal of exposure, both in cells . . . and in humans" (Goodman et al. 1995:288–89). Or perhaps the biological effects may stem from processes that respond only to a narrow range of voltages or currents (Becker and Selden 1985:143, 176–77) or frequencies, for example, those involving chemical species containing "unpaired electrons," say, the free radicals that are generally believed to play a central role in aging processes (Walleczek 1995), or those controlling signaling across membranes (Luben 1995), or those managing melatonin production (Reiter 1995); such specificities could defy many trial-and-error attempts at replication.

So in any matter where the variables are numerous, the unknown unknown may hide vital clues for quite a long time, centuries in the case of bioelectromagnetism. To contemporary science and contemporary common sense, research might look foolish or pseudoscientific even when it is on the track of some important truths about nature. The lack of reproducibility after a century or so of parapsychological research is often cited as proof that there's no real phenomenon of *psi.* Those who hold such a view, however, might chew on the fact that unquestionably genuine electromagnetic phenomena in living systems are slowly being elucidated only after several centuries of irreproducibility.

It isn't so easy to reproduce exactly what others have done even if they publish apparently complete descriptions: "those attempting replication should spend time in the original investigator's laboratory to understand all the subtleties of the experiment, prior to attempting replication in their own laboratory" (Goodman et al. 1995:325). That stricture comes not from parapsychologists or cold-fusion buffs but from researchers at the Biomedical Research Institute of the University of Wisconsin and the Office of Naval Research.

Though the systems in which cold fusion has been reported may seem a little less complex than living systems, nevertheless the number of variables that might influence such a process is very great; and in both cases one type of possible theoretical explanation draws on conceivable "coherence" characteristics, an area that science is barely beginning to understand.

Reasonable Speculation: Connecting Theories — Until reproducibility is attained, methods remain in flux and their relation to mainstream methods is moot. So too with theorizing. Until we understand a phenomenon rather fully, we cannot know whether the reasoning that led to its discovery was inspired prescience or the purest serendipity. Greek science theorized about geometric and arithmetic matters and sought clues to the composition of matter in what we now know to be inappropriate "numerological" directions; in medieval times when theorizing naturally focused on "shape-determining" factors, it was reasonable to class together sharks' teeth and arrowheads.

Before there's a reproducible, well-characterized phenomenon, theorizing is really a matter of speculating. History demonstrates that only through hindsight can speculation about technical or scientific possibilities be accurately labeled as valid or unsound. Some early ideas about electromagnetic interactions with bone were based on the notion that application of a field analogous to that displayed by bone under normal stress, or analogous to wound or injury currents or voltages, could effect healing; and successful treatments have ensued. Does that mean the reasoning was entirely sound? Could one not have likened it to the homeopathic notion that whatever causes symptoms of a particular sort can also, at different dosages, remove the symptoms? Or to the magical "principle of correspondence," which we nowadays reject, under which Mars and war were related through the color red?

Surely it was reasonable to dismiss the possibility that electromagnetic fields could exert any biological effect when their magnitude is small compared to that of electromagnetic fields already present in the living systems themselves. But such effects continued to be observed, as mentioned earlier: "living cells discriminate against random thermally generated EMFs while responding to externally applied fields by recognizing . . . [their] spatial coherence" (Mullins et al. 1995:336).

Was it reasonable to speculate that because accidentally induced insulin shock cured some people of madness, therefore other means of inducing shock, such as electric shock, might cure madness (Rowbottom and Susskind 1984:192)?

Was it not reasonable to dismiss out-of-hand the possibility that humans and other animals could respond to the Earth's magnetism, given that our bodies contain no magnetic organs? Or might one have hesitated on the grounds that we don't yet know everything about our physical makeup?

It's all very well to accept it when something works even if we don't understand how. It's another matter, however, to consider how such a thing could come to be discovered in the first place (see my discussion in chapter 7 concerning Paracelsus and sympathetic magic). What exactly makes speculation reasonable as opposed to unwarranted?

Pseudoscience and Technology

To the wise medical practitioner, medicine is an art based on experience and assisted by science (Nuland 1998). For the philosopher or the pundit, the efficacy of medical treatment is an exquisite mixture of material efficacy—patients getting and feeling better for tangible physical reasons—and "mind over matter"—patients feeling better for no tangible reason (Harrington 1997; Shapiro and Shapiro 1997). But in the long run and summed over all the patients who tried it, a remedy must positively work to remain accepted as legitimate. Whether it's understood *how* it works is of secondary importance. The history of Western medicine is full of treatments that have worked well but whose action was not understood until long after their use had become established—aspirin, for example, or hypnosis.

There's also much leeway in what's meant by "understanding the mechanism" in living systems. The use of pulsed electromagnetic fields (PEMFs) to help in the healing of bone fractures has not only a couple of decades of practical experience to support it but also a connection with known piezoelectric properties of bone; nevertheless, "many authors continue to promote the notion that 'the mechanism' is unknown without defining whether they are referring to the biological or the physical effects of PEMFs." A portentous consequence is that "although . . . [regulators] approved PEMFs as safe and effective in 1979, [they] continue to demand an indication-by-indication approval for any new application" (Bassett 1995:266).

So one cannot automatically apply to medicine the suggested "connectedness" criteria for discriminating science from pseudoscience: successful medical treatments may have no evident connection with existing theory; the *method,* the procedure, may be entirely novel; it's only the *facts* that count—do patients recover or not? In basic science, radical novelty in two of the triad of facts, methods, and theory would at best qualify as "premature science" that typically remains largely ignored by most of the mainstream community; or, if the facts of patient recovery were not overwhelmingly compelling, such radical departure from the conventional wisdom would in science be called pathological or pseudo. In this respect, medicine is quite like other technologies: efficacy is all that counts. Medical advances often depend on adventurous practitioners and brave (or unwitting) patients; an example is cauterization by electromagnetic means to treat aneurysm (Rowbottom and Susskind 1984:144–45). X rays were used medically from the time of their discovery, both therapeutically (to destroy cancerous tumors, for instance) and for seeing bones. On the other hand, reliance on experience can also cause harm even if it is accepted orthodoxy, for the experience may be fallacious. For instance, conventional medicine for much of the eighteenth and nineteenth centuries subjected patients to "heroic" treatments like bleeding that we now know to have been harmful rather than helpful.

Robert Becker "found that engineers were often more open-minded than biologists" (Becker and Selden 1985:99). Technologies—engineering, medicine, art—are on the whole more open to revolutionary proposals than is science, because they are not hamstrung by the need for orthodoxy in method and theory as well as in facts. But the other side of that coin is that technology and medicine are more gullible toward extraordinary claims.

Recurrences

The history given earlier in this section is replete with déjà vu: rediscoveries of phenomena—some of them almost certainly genuine, others still doubtful—and also commercial exploitation based on claims that bear only a certain vague conceptual relationship to frontier research being carried on by entirely competent and respectable people. The contemporary advertisements for magnetic insoles, pillows, and the like mimic those of 150 years ago. Burr's "L-field" of the 1920s and Becker's pattern of DC potentials that evidence an holistic organismic organization have had vitalist precursors into antiquity.

Not only spurious discoveries or pseudoscience fade away in between periodic recurrences. The indubitable successes of electrical stimulation of bone healing since the 1960s had been anticipated by more than a century: "in 1812, Dr. John Birch of St. Thomas' Hospital in London used electric shocks to heal a nonunion of the tibia. A Dr. Hall of York, Pennsylvania, later used direct current through electroacupuncture needles for the same purpose, and by 1860 Dr. Arthur Garrett of Boston stated in his electrotherapy textbook that, in the few times he'd needed to try it, this method had never failed" (Becker and Selden 1985:172). It's humbling to note that even sound knowledge (or at least sound practice) could come to be abandoned for reasons that, in retrospect, we recognize as unsound.

Archaeoastronomy

What did humankind know of astronomy in prehistoric times, before there was any written record that we can now examine? What was the cultural significance of that knowledge?

Archaeoastronomy addresses those questions. Answers have been offered by an eclectic mixture of people from within and without a variety of academic disciplines. Introductory surveys especially worth consulting are by Aveni (1989) and by Krupp (1978a, 1997).

Krupp's book was touted in a jacket blurb as the "first popular over-all survey of the exciting new science *Archaeoastronomy*" (Krupp 1978a). The author pointed out in his text that "across the whole face of the earth are found mysterious ruins of ancient monuments . . . with astronomical significance. . . . Some of them were built according to celestial alignments; others were actually precision astronomical observatories. . . . We are just awakening to the success achieved in astronomy by many . . . prehistoric peoples. . . . Now, in an era of wild, unsubstantiated claims about ancient astronauts by Erich von Däniken and his fellow travelers, their pseudoscientific misconceptions of these earlier peoples are in part dispelled by the reliable, scientific findings of archaeoastronomy" (Krupp 1978a:xiii).

Britain and other parts of Europe are studded with "megaliths"—stones, often but not always large, and arranged in circles, or in linear fashion, or as single "outliers," or as "trilithons" rather like doorways, and in other ways as well. They were built over some thousands of years, earlier than about the second millennium BC. Associated with these enigmatic constructions are burial cairns, passage graves, and elaborate mausoleums like Maes Howe on Orkney and Newgrange in Ireland, where at the winter solstice the sun reaches and illuminates the inside as on no other occasion.

These marvels were built by people who left no written records and who were, according to conventional archaeological wisdom, subsistence farmers, fishers, and hunter-gatherers living in scattered groups of extended-family size. Yet the building of those megalithic wonders called for extraordinary organization and huge expenditure of effort, not to speak of architectural skill and astronomical knowledge.

The difficulty of trying to infer what the social arrangements were of preliterate people thousands of years ago has inevitably bred controversy; "the debate between astronomers and archaeologists . . . initiated archaeoastronomy as an interdisciplinary field of study" (Baity 1973:390). The mainstream often rejects at first what to outsiders may seem eminently reasonable speculation, for example, Euan Mackie's about social differentiation (Mackie 1977a,b) or John North's (1996) over astronomical alignments. Stonehenge is the site most argued on this score (Krupp 1978b).

Speculation shades into the fantastic as megalithic sites are alleged to show some overall pattern of connections via "ley lines," straight lines connecting three or more significant sites (Krupp 1978c); some go even further to suggest that these alignments represent actual "power" lines (Michell 1975). There is an enormous literature ranging from staidly descriptive through speculative to fanciful about the Egyptian and Mesoamerican pyramids as well as the European megaliths.

As with bioelectromagnetics

- Only in recent decades has the recognition become at all general that there is a body of sound though scattered knowledge awaiting and perhaps ready for integration.
- There is little of the institutional infrastructure that established disciplines have. There are journals and societies and meetings, but there are no degrees or departments or individual positions allocated specifically to this field of study.
- Participants come from a great range of disciplinary backgrounds.
- Controversy is routine, over quite fundamental issues.
- Participants are not always enamored of the credentials of, and approaches used by, some of their bedfellows.
- Some of what is written on the subject can properly be called cranky or pseudoscience.

Like bioelectromagnetics, then, archaeoastronomy is in a number of ways more akin to anomalistics than to an established natural or social science, even as it is practiced largely by people schooled in traditional disciplines.

Overlap of Fringe with Mainstream

Pseudoscience in archaeoastronomy is exemplified not only by Erich von Däniken and ley-line speculators but by a host of other people and topics: the Glastonbury zodiac, the Sirius "mystery," Atlantean traditions in ancient Britain, correlation of flying saucers with ley lines—a veritable New Age feast. The Great Pyramid of Egypt has inspired a wide range of esoteric notions; C. P. Smyth, a Scottish astronomer, in the mid-nineteenth century claimed to find in its dimensions insight into a number of facts about the universe—which may well have prejudiced archaeologists for decades thereafter against astronomical interpretations of ancient sites.

Perhaps the longest-running focus of wide-ranging speculation has been Stonehenge, which is unique among stone circles in several respects, most obviously the exact shaping of the stones. Such otherwise respected astronomers as Gerald Hawkins and Fred Hoyle have suggested that Stonehenge functioned as (among other things) an eclipse predictor and that it was built at just the right latitude to make its geometry spectacularly symmetrical. Some archaeologists have been unkind in their critiques of such suggestions, which are however staid in comparison to popular myths attached to Stonehenge about Druids and human sacrifices.

Connectedness

In *method* and *theory*, archaeoastronomy suffers the typical dilemmas of interdisciplinary fields. Participants come from a variety of backgrounds and need to learn how to adjust to and incorporate ideas and approaches from other disciplines.

The *facts* of archaeoastronomy are a source of considerable and inevitable controversy. The direction of any purported alignment can be (and generally is) argued over because

- There is no unambiguous line. Alignments depend on notches along the horizon, stones or pairs of stones as foresights, gaps between stones—all of which are subject to significant uncertainties. There are no unequivocal, exact directions.
- The marker stones and buildings may no longer be in their original positions. Many have fallen or been damaged or moved. For instance, a "sun dagger" in the U.S. Southwest was identified only to be lost through a slight shift in a stone slab: "skeptics will doubtless note the irony that after some 1000 years of accurate patterns, the rocks shifted their position barely a decade after they were discovered" (Palca 1989). Had that "alignment" then been just a fluke?
- Where the horizon is part of the alignment, trees and other vegetation now present may not have been there earlier; but also those that may have been present in earlier times may not be there now.
- The significance of an alignment depends on the era. If the dating of a monument is changed by a few centuries, that can change the astronomical event that alignment supposedly indicates.
- If there are many possible astronomical features, then it becomes likely that some alignment will be indicated at some date or other, by sheer chance. This point becomes salient as alignments are looked for not merely with significant aspects of sun and moon but also with the rising or setting of various stars; the more so as "rising" and "setting" are affected by the distance above the horizon that is taken as the key point—especially when the "horizon" is said to have been defined by a ditch or an earth-bank whose original depth or height can only be estimated (North 1996).

Even when no precise alignments are asserted there can still be controversy.

No one denies that looking toward the avenue at Stonehenge from the circle's center points generally to the rising sun at the midsummer solstice. Yet that may not have been the builders' concern—perhaps the significant line of sight was in the opposite direction, from the avenue toward the ring at midwinter (North 1996).

As to precision, there is dispute about how accurately the prehistoric observations might have been made. Alexander and Archibald S. Thom, a father-and-son team of engineers, are universally credited for making measurements of unparalleled reliability at a host of sites; yet their claim is far from universally accepted that the constructions reveal a prehistoric measuring unit, the megalithic yard (MY); still less accepted is its claimed accuracy of 2.720 ± 0.00 feet.

Interdisciplinarity

Archaeoastronomy draws centrally on astronomy and archaeology. Engineering is needed to understand how the impressive ancient monuments might have been constructed. Linguists and specialists in cave and rock and ceramic art have valuable perspectives and insights to share. Like archaeology, archaeoastronomy has to consider suggested interpretations of folklore, myth, and the earliest written records—another similarity with anomalist fields like cryptozoology and ufology. For instance: is it coincidental that Diodorus reports visits by Apollo to his "spherical temple" served by the Northerners (Boreadae); and that at Callanish in the Outer Hebrides in megalithic times "the moon was for a few days circumpolar, never setting at all" (Baity 1973:397n)?

This new field requires both a disciplinary merging and a breaking down of traditional subfields. British and French sites show analogies and similarities, and the Thoms have suggested that the measuring unit in Brittany was identical with the British rod. Comparisons need to be made also with Middle Eastern and Asian sites, if not with American ones as well.

The typical problem of interdisciplinary and anomalistic fields was pointed out in an early review, that "no indexing system exists" (Baity 1973:390)—no way of locating essential literature in the many relevant disciplines. Precursors of work in archaeoastronomy have progressively come to light (Baity 1973:394).

Not only professionals from various disciplines but amateurs quite outside academe have done some of the groundbreaking work, for example at Stonehenge by C. A. Newham, who was not paid much attention at first by archaeology. One of the early North American Indian studies, of the solstice-indicating "sun dagger" at Chaco Canyon mentioned above, was made by an artist

Anna Sofaer. A local archaeologist declined to visit the site, and two archaeoastronomers did not take her claim seriously; but eventually she was able to publish, with an architect and a physicist as coauthors (Sofaer et al. 1979).

Archaeology did not embrace even professional astronomers who wanted to interpret archaeology's data without having first been steeped in it: "the astronomers and the archaeologists are not speaking the same language"; "a brilliant and explosive theory has gone rocketing ahead without anything like critical control" (Baity 1973:400). Speculation about Stonehenge by the astronomers Hawkins and Hoyle was described by the leading archaeological authority on Stonehenge as "Moonshine on Stonehenge": "tendentious, arrogant, slipshod, and unconvincing" (Krupp 1978b:104). Another leading archaeologist did not dismiss the astronomy but regarded it as of small interest (Krupp 1978b:110).

The astronomers were willing to presume that an alignment was an alignment and therefore significant, no matter how unlikely that might seem from what archaeologists then believed about the way of life of the people concerned—who after all had left no written records, had lived in primitive stone-and-wattle houses, and had used stone and bone tools. Outsiders to archaeology, on the other hand, might find it strange that the insiders accepted that such a primitive, still "partly nomadic society" (Krupp 1978b:86) could put up such massive artifacts as Stonehenge, Avebury, or Silbury Hill—the latter the largest artificial mound in Europe, 550 feet across the base and 130 feet high, estimated to take five hundred men ten years to complete (Reader's Digest Association 1977:316).

If archaeoastronomy is at all right, then "we must alter radically our . . . view of the intellectual calibre of man in Britain in the late third and second millennium BC" (Baity 1973:415).

What Is Relevant?

In anomalistics, it is not always clear which phenomena are related to one another. In ufology, for instance, there are sighting reports and mysterious radar echoes and alleged physical traces or electromagnetic effects and stories of being abducted and disparate photos and films; do all of those stem from a single phenomenon? So too in archaeoastronomy: is there a connection between astronomical features and various symbols in cave paintings and stone etchings? To outsiders many other explanations might occur than that primitive spiral shapes or cup-marks are representations of things in the sky, or that "tallies" of nineteen marks should be interpreted as representing years in some eclipse cycles or 29½ marks the days in a lunar cycle.

Much to Be Done

Archaeoastronomy as a self-conscious interdisciplinary endeavor dates perhaps only to the 1970s. But individuals had made such studies much earlier. At the end of the nineteenth century the astronomer Norman Lockyer recognized astronomical alignments in Egyptian and European megalithic architecture. The Thoms have made site measurements and offered interpretations since the 1930s. The review article published in 1973 by Elizabeth Chesley Baity, "Archaeoastronomy and Ethnoastronomy So Far" (Baity 1973), and the appended commentaries from people in a variety of disciplines convey a sense that a new synthesizing venture was being born. The Royal Society and the British Academy held the first meeting on ancient and prehistoric astronomy in 1972. The journal *Archaeoastronomy* was inaugurated in 1979 (Lankford 1997:21–30).

Archaeoastronomy, like bioelectromagnetics, is not yet a coherent field in which there is consensus over where the major lines of research ought to be directed; rather, small enthusiastic groups work at various aspects of it. Most of the work has been done around Europe. Just the last few decades have seen dawning realization that North American Indian sites incorporate astronomical features. Central American and South American monuments have barely been explored let alone fully documented or analyzed. (E. C. Krupp commented in April 1999, however, that the best archaeoastronomical work is having an impact on mainstream archaeology and anthropology, and those fields are increasingly taking account of astronomical and cosmological issues.)

That so much remains little known leaves the subject rather defenseless against the poorly based speculations of people like von Däniken. At the same time the field draws attention to fascinating questions in human prehistory: given that megalithic astronomy had achieved so much, and had surely been useful in agriculture and social organization, why did these sites fall into disuse?

Other Borderland Subjects

Bioelectromagnetics and archaeoastronomy show how diffuse and porous can be the borders between mainstream and unorthodoxy on subjects that are not yet established disciplines. Rather than insisting on a stark dichotomy between genuine science and pathological or pseudoscience, one must recognize that quite a number of pioneering studies straddle the fence. For one who becomes interested in inquiring into the interstices between what the established fields know, the following few illustrations underscore that there is no scarcity of fascinating topics to delve into.

The Body

Clinical Ecology — Clinical ecology holds that some people suffer drastically through exposure to substances that most other people tolerate without difficulty. For example, some people are said to be so allergic to wheat as to suffer the most severe arthritic symptoms—which can however be reversed almost instantaneously when wheat is avoided. This is just the sort of claim that's dismissed by many experts as on-the-face-of-it pseudoscience, even when it is illustrated by an actual filmed demonstration like the one I saw sometime in the 1980s on British television. Yet "bad building syndrome" and "chemical sensitivity" have become accepted as real, leaving between mainstream and unorthodoxy a rather fuzzy region (Hanson 1990). The distinction may be chiefly in the mainstream's unwillingness to acknowledge the success clinical ecologists seem to have had in treating a variety of patients (Ashford and Miller 1990; Hileman 1991).

Chemical sensitivity "illustrates the classical way in American medicine of dealing with diseases of unknown mechanism. . . . Before the role of prostaglandins in primary dysmenhorrea was described and prostaglandin-inhibiting drugs developed, women with this disease were treated as psychiatric cases. . . . Both the physician unable to treat primary dysmenhorrea and the industrial hygienist have an interest in maintaining the status quo, if only to save face" (Morrill 1990).

Physique and Psyche — Many correlations or cause-and-effect relations between physiology and psychological states have been asserted or believed at various times. Among those that we now regard as pseudo is palmistry. Surely there's no way that the lines on one's hands reflect anything about one's mental or emotional experience, let alone what the future holds. Phrenology, by which the landscape of the skull was supposed to reveal intellectual and artistic abilities, is now classed with N rays as a classic instance of pseudoscience. Then there was a theory of criminal types: body and facial structure could supposedly identify criminals.

Yet even as we snigger at such outdated superstitions we take seriously others. For instance, in biofeedback, through observing our own blood pressure, pulse rate, and the like, we can learn to control them. Intelligence (or at least a measure of it like IQ) is believed to have a genetic component. Violent behavior may have a genetic basis (Duster 1992).

Are these latter body-mind relations really so obviously plausible by comparison with the supposedly superstitious ones?

What's Inside Cells? — "That the nucleus runs the cell seems obvious. But when cells are gently stained the elusive nucleolus appears—and seems, at times, to be in charge" (Horrobin 1981:266). This is a heretical notion that brings outraged responses (Ashburner 1981a,b) as well as interested ones (Arnold 1981).

The claim that cell features seen under electron microscopes are artifacts resulting from the method of preparing specimens, which seems probable rather than merely plausible, was banished from mainstream thought (Hillman 1987a,b, 1988; Wilkinson 1988).

Until Prusiner's Nobel Prize was announced, I had intended to mention prions here rather than earlier. HIV as the cause of AIDS would fit here as well, as would the question of the origin of life.

The Earth

Gaia — "[M]any earth scientists . . . [have come] to think about the biosphere not as a bunch of 'coevolving organic and inorganic parts but rather as a single entity made up of living organisms: an atmosphere, an ocean, and a planetary surface all interacting with one another through a series of complex feedback processes'" (Aveni 1992:214–16). This is "the Gaia hypothesis. . . . Is the world more than the sum of its parts? Does it possess a purposive element? Is it conscious? The theory of Gaia seems to be part of the drive to rehitch culture to nature. Milder in form and positioned as it is in the penumbra of the many millennial offspring of its occult sisterhood of astrology, it nonetheless invites open conflict with scientific catechism" (Aveni 1996:301–2).

"All life on earth can be considered a unit, a glaze of sentience spread thinly over the crust . . . a hollow, invisible sphere inscribed with a tracery of all the thoughts and emotions of all creatures," like Teilhard de Chardin's postulate of a noosphere, or ocean of mind (Becker and Selden 1985:269–70).

Earthquake Prediction — Near panic ensued in the Mississippi Valley at the prediction that a major earthquake would strike there in the first week of December 1990 (Kerr 1991). The prophet, Iben Browning, was credited in the *New York Times* with having predicted a week in advance the San Francisco earthquake of 1989. To the media, Browning was a competent scientist—though his Ph.D. was in zoology, not geology or seismology. He did also have the support of a seismological insider, the director of the Earthquake Information Center at Southeast Missouri Sate University.

There's disagreement within the mainstream community as to whether earthquake prediction is possible (Geller et al. 1997). However, that is a disagreement over the possibility of *specific* predictions; there is no dispute that zones can be identified in which earthquakes are relatively likely, and other regions in which they're extremely unlikely. One of the ironies of the Browning episode is that the Midwest *is* prone to rare but lethal quakes, contrary to the public belief that such events are restricted (in the United States) to California (Kerr 1991).

Water from Comets — Louis Frank is sufficiently distinguished a physicist to be awarded a "named chair"—he is the Carver–Van Allen Professor of Physics at the University of Iowa. In 1986 he discovered that minicomets, composed chiefly of water, are hitting the Earth at an astounding rate: they could account for a large proportion of the water presently on Earth. The discovery was soon pronounced spurious by most of Frank's colleagues (Frank and Huyghe 1990). Nevertheless, more than a decade later there are indications that Frank may yet be vindicated.

History and Culture

- What *did* happen to Czar Nicholas and his family? To Martin Bormann? To Jimmy Hoffa?
- Who wrote Shakespeare's plays? (Lardner 1988; Sobran 1997)
- Who was the "Man in the Iron Mask"?

These and many similar questions occupy both orthodox and unorthodox researchers and pundits. The chief facts of a matter may not be disputed while their interpretation is, for example, over the assassination of Martin Luther King or the U.S. government's attack on the Branch Davidians at Waco, Texas. Factual details may be argued over much as they are in scientific anomalistics—identification of bullets and the like—but the essence of the dispute usually is a clash of ideologies more than a dispute between simple knowledge-seekers.

First in America — The dogma is only slowly dissolving away, that the first humans in the Americas were those who left their mark in the form of Clovis points, not much more than twelve thousand years ago. South American sites *somewhat* older seem recently to have attained wide acceptance, though skepticism remains strong over *much* older sites excavated by the same competent people in the same region—sites that appear to be on the order of thirty-five thousand years old (Meltzer 1997). Yet there is really no good reason in logic or scientific knowledge to dismiss that claim.

Evidence from genetic and linguistic studies taken together with the newly accepted archaeological data appears ready to bolster the revision of the long-standing official consensus that the Americas were settled in one or two migrations across the Bering Strait. It now seems that there were perhaps four major groups of people involved (McDonald 1998). Perhaps there was even some trans-Pacific settling of South America; a skull more than eleven thousand years old found in Brazil appears anatomically kin to those of South Pacific islanders (Anon. 1998c).

Unorthodox and far-out suggestions about the prehistory of the Americas are legion (Dixon 1993; Huyghe 1992).

Animal Language — Can animals (apes, birds, and others) be taught to communicate in quasi-human language, if not by sound then by signing? Some experts say yes, others say no, and the latter sometimes call the very suggestion pseudoscience. At the same time it is widely accepted that ants, bees, birds, dolphins, whales, and other species are able to share quite specific pieces of information among themselves.

6 Scientific Heresies: Beyond the Pale

Most of the popular or populist preoccupations so often lumped together as "pseudoscience"—ufology, pyramid power, astrology, and so on—exist almost entirely outside scientific communities. I am suggesting that the activities of those who pursue these popular subjects can be divided into two distinct categories:

1. Genuine knowledge-seeking about matters ignored by the established intellectual disciplines, in other words anomalistics. *The prime motive is intellectual curiosity.*
2. Self-promotion through dogmatic insistence on palpably unproven points. *The prime motive is personal gain.*

Only the second deserves to be labeled pseudo.

Of course it is a risky business to make distinctions based on motive. And of course the first category shades into the second. Those who seek unorthodox knowledge often suffer conflict of interest because they need support and are tempted to oversell their wares. But that same conflict of interest afflicts those in the mainstream also. Again as in the mainstream, knowledge seekers often become convinced they are right long before others are prepared to agree. Nevertheless, some such distinction is needed to avoid the unsound lumping together of topics that almost beyond doubt have no redeeming grain of truth with those pointing to possibilities that may lurk in the unknown unknown.

My proposed distinction also has a practical side: one can more readily distinguish dishonesty from honesty than ascertain the truth about claimed phenomena that science does not fully comprehend.

Commercial Huckstering

Some of the topics frequently called pseudoscience really are that, from any number of perspectives. The data may be spurious to start with, the presentation deliberately misleading, the proponents' purpose to gain fame or fortune from the unsuspecting gullible public—nothing about the subject bears comparison with genuine knowledge-seeking. It's easy to sympathize with debunkers who get angry about the prevalence, in newspapers and magazines with mass circulation and in nationally broadcast media, of the sort of rubbish that appeared in several issues of *TV Guide* in 1998, on a page headed "TV GUIDE MART—ATTENTION ADVERTISERS—Reach over 22 million responsive readers":

> **"Astrology/Psychics"—**
> LOVE, MONEY, SUCCESS, find out what's in your future. Call our gifted psychics at 1-800-689-TELL . . . Adults over 18. 24hrs.
>
> REAL PSYCHICS! Don't be fooled by imitations. Call the American Association of Professional Psychics®. Registered with the U.S. Government. Guaranteed Authentic. 1–800/810–2266.

Such things are less instances of pseudoscience than they are of fraudulent marketing. After all the same page offers "**CARS FOR $100!** Seized & sold locally. Porsches, BMW's, Corvettes, 4x4's, motor-cycles, boats, etc. 1–800/522–2730 Ex.2362"; and "**BUSINESS OPPORTUNITIES—GUARANTEED PROFIT!!** Start Your Own Internet Business. No Computer Necessary. Not Multi-level, $200 Start-up. 1–888/418–6677."

Offers of healing and health in this genre abound. This example comes from a well-established mail-order house:

> **Magnetic patches** apply the ancient Chinese principles of magnetism to help the body heal itself. . . . Set of 20 dots and patches **$7.98. . . . Magnetic cervical collar:** For centuries, Chinese healers have used magnets to relieve pain and reduce swelling, soreness and stiffness . . . **$9.98. Magnetic pillow** gives you the most restful, restorative 40 winks of your life! Ancient beliefs hold that strategically placed magnets may help relieve pain and stress. . . . You'll wake up more relaxed and revitalized . . . **$12.98, Two for $24.50.** (Carter n.d.)

Similarly, an in-flight mail-order catalog advertised "**Biomagnetic Pain Relief** . . . Magnetic Field Therapy. Bio Magnetic International products' magnets help attract oxygen-rich blood . . . $39.95" (*SkyMall* 1998).

Magnetism is peddled not only for these powerful biological effects unknown to science but also for powerful *physical* effects unknown to science. According to some claims, the efficiency of automobile engines can be increased if magnets are mounted around the fuel line, say, the Tripolian at $55, touted in the Whitehorse Research Catalog (at <http://whitehorse-research.com/l/l2.htm> in 1998), or the Optimiser for diesel engines (Howard 1991) or a new magnetic fuel-conditioning device for cars that "rearranges fuel molecules and enhances the bonding force between hydrogen and oxygen in the fuel mixture" (Reese 1992). Such charlatanry reveals itself by scientifically meaningless—not to say absurd—phrases, such as "magnetically ionizes," "rearranges fuel molecules," or "enhances the bonding force."

Other companies claim that water-heaters will not build up scale if magnets are allowed to perform their magic on the water: "**Mineral Magnet—The scale & deposit eliminator . . . Does not alter water's vital minerals . . . $39.99**" (*Bright-Life* 1997:14); "Taste, see and feel what a difference cleaner water pipes make. The amazing Mineral Magnet . . . $39.99" (*Bright-Life* 1997:30); "Hard Water Conditioner . . . PVC and ceramic ferrite magnets. Works for at least ten years. Doesn't affect taste or health benefits of water . . . $39" (*Handsome Rewards* 1996–97, item 2599).

A single product can do both jobs, of course. "SoPhTec International designs and manufactures highly effective physical (magnetic) water conditioners. . . . There is also extensive use of the products for fuel conditioning on both private and commercial vehicles," according to a Web site advertisement from SoPhTec International, based in Costa Mesa, California (<http://generalenv.com/SoPhTec.html>). The smaller C-250 unit can be had for only $90; the larger C-500 costs $269.

Moreover there are health benefits to the water-treatment: "Conventional water softeners attack the problems of hard water by replacing beneficial calcium with unhealthy sodium. This can lead to high blood pressure and heart disease. It also pollutes groundwater. Physical (magnetic) water conditioning, on the other hand, is superior because it solves the problems of hard water without adding or removing anything from the water or creating pollution," according to the Web site of General Environmental Systems, Inc., of Tucson, Arizona (http://www.generalenv.com; water@generalenv.com).

These are no appeals to knowledge seekers. It makes no sense to confuse this sort of fraud with empirical and theoretical arguments over ufology, cryptozoology, parapsychology. Scientific terminology—"magnetic"—is being deliberately used to obfuscate, deceive, and exploit.

Mainstream commercial advertising does the same thing all the time without being labeled pseudoscience: Quaker State Oil "*with QSX,*" say, or dietary supplements "with minerals and vitamins." "QSX" and "minerals and vitamins" function like "abracadabra." The audience is not expected to understand what they mean but to grant them power in the faith that the shamans do understand.

Given that science has in fact delivered so many marvels that the general public doesn't understand, there's no easy way to counter this trash. To stem its misleading flow calls for action on educational, political, and legal fronts. There is little if any intellectual content to argue over, no issue of what's science and what isn't.

Grains of Truth

That hucksters sell magnetic or other quack remedies of no proven efficacy does not however establish that there is no effect of magnetic fields on living systems. That there is no known physical basis for magnetic water or fuel treatments does not exclude the possibility of some as-yet-unknown phenomenon related in some fashion to such claims.

There are in fact a few people who seem concerned to investigate rather than to sell magnetic devices. Conflicting results on magnetic water-treatments continue to be reported (Powell 1998). "There is . . . an extensive literature dealing with the effects produced on water by magnetic fields" (Franks 1981:46). Still, even some who have observed an effect acknowledge that "many magnetic devices on the market 'are completely ineffective'" (Reese 1996). The possible influence of magnetic fields on chemical reactions remains an open question (Atkins and Lambert 1975; Barron 1994) though it has "long been a romping ground for charlatans" (Atkins 1976:214).

Even some of the worst commercial huckstering that trades on people's gullibility can at least provoke awareness of how much we still have to learn about all sorts of things. We can brush this stuff aside and remain no wiser than before; or we can ask, "What *exactly* do we know about magnetism that makes this claim insupportable?"—and then we just might learn something new. The future quite likely holds further discoveries connecting magnetism and the behavior of living creatures (Livingston 1998). The mainstream literature includes such tantalizing albeit scattered tidbits as the following:

- There are magnetite crystals inside humans as well as in other animals and also in bacteria (Baum 1992; Corliss 1994:114, 136, 143, 176, 208).
- Humans, homing pigeons, and some other species seem to have a directional sense guided by the Earth's magnetic field (Baker 1980; Corliss 1994:114–15, 143, 167; Kalmijn 1982; Phillips 1986).
- Dowsing successes seem to correlate with magnetic anomalies related to the presence of the water (Corliss 1994:303; Williamson 1987).

- The occasional stranding of whales may correlate with anomalies in the Earth's magnetic field in the region of the shore (Corliss 1994:135–36; Kirshvink 1984).

Some *electro*magnetic effects on living things have become quite well established, as discussed earlier.

Wellness Quackery

Claims about physical effects of magnetism can be tested in fairly straightforward ways. But promises that magnets can make people *feel* better are much more difficult to check. Assertions about health are probably the most prevalent form of commercial, huckstered pseudoscience. Los Angeles, California, may be unsurpassed in this. Within a few hours there I picked up half-a-dozen free magazines and newspapers offering

- unsupported assertions that exploit wishful thinking: "Improve your life and the health of our planet" (*Awareness,* September–October 1997:1).
- combinations of unorthodoxies: "Ghosts, Spirits, UFOs . . . lecture & channeling session led by nationally known psychic" (*The Whole Person:*26).
- omnivorous combinations: At the Krause Holistic Healthcare Center, take advantage of "Homeopathy, Activator, Light Force, Diversified Network Chiropractic . . . Shiatsu . . . Micro-Current Therapy . . . Iridology, Kinesiology" (*Awareness:* 19). Available at the Kastin Healing Arts Center are "Colon Hydrotherapy" and "Lymphatic Drainage Massage" (*The Whole Person:*5). "The Institute of the Healing Arts (a state licensed and CCPVE approved school)" offers some of those and also "Acupressure, Alexander Technique, Animal Communication/Healing, Aromatherapy, Aura and Chakra Balancing . . . Ear Candling, Feng Shui, Flower Essence . . . Holographic Repatterning, Reflexology . . . Tai Chi, Tarot, Trager, Tui Na" (*The Whole Person:*29). Other establishments offer various combinations of some of those with others again such as CranioSacral Therapy, Ear Coning, Esalen, Genetic Regression, Johrei (purifies the Spirit through the channeling of Divine Light), Polarity, Psycho-Cellular Healing, Reiki, Rolfing, Shamanic Mentoring (including "Soul Retrieval"), Rebirthing, Sonic Empowerment . . . Shadow Tracing for Upgrading Your Parallel Lives.

- mumbo-jumbo that sounds scientific without actually being so:

  "pyruvate and a variety of other dieting programs" (*New Times*, 1997);

  "in the periodic table there were actually elements that consisted of a single atom per molecule. . . . [T]hese . . . could actually generate photons of light and within the body this transmission of light might allow illness to be healed. . . . [T]he complex sugar molecule (called mucopolysaccharide . . .) . . . contained monatomic elements" (Wheeler 1997);

  "Another way shamanic thinking is infiltrating modern medicine is [through] . . . systems dynamics computer modeling that predicts future illness and disease in individuals. 'It's based on chaos theory, simultaneous non-linear differential equations . . . a kind of mathematics that uses imaginary numbers and is very effective for looking at phenomena for which there are multiple factors, all of which are interrelated, which is what happens in health and disease; there's no independent variable, only loops and feedback'" (Anon. 1997b).

Again I suggest that what is offensive about this is not so much its lack of intellectual validity, not that it's pseudoscience, as that it is deliberate deception for personal profit. Immanuel Velikovsky or Wilhelm Reich may have been just as wrong in some of their assertions, but at least they were honestly misguided and were genuinely seeking knowledge, not deliberately or recklessly setting out to mislead others and gain therefrom. I think that difference is an important one.

Velikovsky

The Velikovsky affair (Bauer 1984a) is as well known to historians and sociologists and philosophers of science as any controversy in anomalistics. A psychiatrist by profession, Immanuel Velikovsky in 1950 published *Worlds in Collision*, which interpreted biblical accounts, myths, and folklore as showing Venus to have emerged out of Jupiter as a comet and approached Mars and Earth a number of times, two or three thousand years ago, with catastrophic psychological (Velikovsky 1982) as well as physical consequences.

Worlds in Collision was favorably reviewed in popular media and became a major bestseller, though it was roundly denounced in scientific circles. A decade later there was further public furor about it when the *American Behavioral Scientist* asserted that Velikovsky had not been treated with proper respect by the scientific Establishment even though his predictions about Venus's temperature, Jupiter's radio signals, and the Earth's magnetic envelope had been verified.

Velikovskian groups and journals flourished through the 1970s. Following Velikovsky's death in 1979, a few groups have continued his "catastrophist" approach to planetary movements and the interpretation of myth and history.

Connectedness

Velikovsky's *facts* were not accepted facts in physical science, though some of them may have been in history, religion, literature, folklore, and the like. His *method* was based not on scientific approaches nor on accepted scholarly ones in history but rather was consonant with his psychoanalytic background. Just as psychoanalysts infer what actually happened from what patients say while free-associating, so Velikovsky set out to infer what really happened on Earth from what folklore and myth and old documents say. Velikovsky's *theory,* in particular his rejection of gravity in favor of electromagnetism (Velikovsky 1946), had no counterpart in physical science.

Velikovsky was not a member of a scientific community. He did make approaches to various prominent scientists, asking that they comment on his ideas or perform experiments that might test them; but he did that as a clear outsider. So far as "connectedness" is concerned, Velikovsky qualifies as a total outsider, disconnected from science and from historical scholarship on both intellectual and social grounds.

Illustrative Points

My book-length discussion of the matter (Bauer 1984a) illustrates points that are of interest for the present purpose. Some of these follow in the sections below.

Fraud — Velikovsky's propositions can hardly be equated with popular trash like newspaper horoscopes or healing magnets. Velikovsky and his followers were and are knowledge seekers, not conscious charlatans or confidence tricksters. They were mistaken usually and incompetent often, but nevertheless honest. The Velikovsky affair has been essentially free of frauds or hoaxes.

Lack of Progress — Unlike science, the Velikovskian "research program" has not made progress. Neo-Velikovskians still stick with basically the same original givens, that catastrophic interplanetary events shaped mankind's history as recently as the first millennium BC.

Half a century after the affair began, much the same points are argued, albeit with different details. Active neo-Velikovskian researchers do the same sort of thing as before: interpreting folklore and myths and calculating speculated planetary orbits, not infrequently in ways that physical science holds to be unsound.

Quality Control — The printed Velikovskian literature lacks quality control. The Internet carries much unreliable stuff by proponents as well as a little counterbalance from "skeptics."

For more than thirty years nobody in mainstream scholarship or science bothered to publish a full analysis and critique of Velikovsky's assertions.

Recurrence — The *recurrence* of anomalies is well illustrated; Velikovsky's notions had all been proposed before, sometimes long before (Bauer 1984a:215–23).

Velikovsky's catastrophism had been argued over for years before it became generally known how similar his work was to that of Ignatius Donnelly half a century earlier, and a host of other precursors came to light as time went by. Devastation on Earth by a comet is a periodically recurring idea, as are millennial notions of impending catastrophe (De Camp 1976).

Role Played by the Great Ones; Schisms among Followers — Without Velikovsky, nothing like his "science" would have come about. While he was alive, his charisma enforced a high degree of agreement. Increasingly now there are different camps disagreeing with one another (and also with Velikovsky's views) to varying degrees on various matters. Some groups focus almost exclusively on reexamining conventional historical chronology, others on scenarios of planetary events; others eclectically attack Establishment science and push Velikovskian science.

Disagreements initially or overtly over intellectual issues in anomalistics can become so bitter that people stop talking to one another or stage legal battles over defamation and the like. Threats of legal action were made by several Velikovskians over the years. Leroy Ellenberger was disinvited from a Velikovskian conference in 1992 because several other participants said they would cancel if Ellenberger were permitted to participate even as a member of the audience.

Amateurs Not Professionals — Velikovskians are primarily amateurs as to science or history. Initially a few—a very few—qualified scientists, historians, and social scientists took Velikovsky's propositions seriously; nowadays there are even fewer if any such.

Velikovsky himself was impressively erudite. Among neo-Velikovskians on the whole the level of scholarly competence and intellectual achievement is markedly lower.

The Appeal Is Intellectual — New recruits to "catastrophism" seem to come largely from a general public that learns of Velikovsky's scenario for the first time, either through publicity by neo-Velikovskian groups or because Velikovsky's books continue to be reprinted; in the 1990s his works were translated into Russian for that newly opened market for anomalistics.

The appeal of Velikovskiana seems to be primarily intellectual, driven by the same sort of curiosity that drives scientific investigation. Personal impressions from a neo-Velikovskian conference in 1992 suggest that some young as well as some older adults who have hitherto had no major intellectual passions find themselves intrigued by the possibility of understanding about science and cosmology and enjoy membership in a community of like-minded enthusiasts. It may be an indictment of our educational system that people who can become so fascinated by cosmology and history and intellectual discussion in general find their curiosity best served by fringe groups rather than mainstream organizations.

Though Velikovsky himself may have had deep religious or ethnic motivation, nothing in Velikovskiana reflects that.

Attracting Support as a Matter of Fairness — A few highly competent people befriended Velikovsky, though it's unclear how much if any credence they gave his propositions. Einstein certainly paid the ideas little heed even as he was courteous toward Velikovsky himself. Harry Hess, a prominent geophysicist, invited Velikovsky to address graduate students in geology. Hess himself had been an early convert to the theory of plate tectonics; he knew that Wegener's notion of drifting continents had been wrongly ignored for three or four decades and so may well have been sensitized to the possible value of listening to far-out ideas; perhaps he was seeking to stimulate open-minded thinking among the students.

The social scientists who supported Velikovsky in the 1960s claimed to do so because of the disrespectful manner in which he had been treated, not because they necessarily took his substantive claims to be correct. Though their discourse was quite critical of how science was being carried on, they held a traditional and respectful view of science and the scientific method as a reliable path to useful knowledge; this is quite distinct from contemporary critiques of science by postmodernists and some schools of sociology (Gross and Levitt 1994).

Grains of Truth, but Wrong in the Main — Velikovsky was not *all* wrong. Some of his assertions turned out to have substance, for instance, that Linear B is Greek, that Venus is hot, that Jupiter emits radio noise. However, the lack of connectedness of those assertions with mainstream knowledge made them useless to mainstream scholarship and irrelevant to science.

Velikovskians try to make these partial or apparent vindications on matters of inconsequential detail grounds for accepting the whole "catastrophist" thesis. Velikovsky's personal charisma and that he was apparently right on such subsidiary points misled quite a few people into taking seriously much of his "science."

Associations — Some competent mainstream scientists with their own unorthodox views, for example, Victor Clube and Tom Van Flandern, accept invitations to neo-Velikovskian meetings. Their ideas being largely ignored or decried by their mainstream peers, they seem to welcome almost any audience willing to hear their case.

Summary

Velikovskian science qualifies as pseudoscience in its lack of connection to the intellectual content of science and to the relevant intellectual communities. Velikovsky does have to his credit being right on some details when the experts were wrong; but his approach is not suited to systematic knowledge-seeking.

But why the passionate denunciations? He was wrong in many ways, but he was not a fraud or deliberate charlatan. He was intensely curious about matters that intelligent people are curious about and went astray largely through going his own way too stubbornly. But he wasn't peddling snake oil or disingenuous tax cuts. The fury of Velikovsky's critics was *odium scholasticum;* in superficial ways he seemed to the public so much like the genuine article.

Wilhelm Reich

It's a curious fact that a reader not already committed in favor of Wilhelm Reich can recognize, from the undisputed facts of his life in books by Reichians, that Reich went intellectually wrong in the 1930s if not before and had become, probably by the late 1940s and certainly well before the end of his life in 1957, socially dysfunctional and intellectually deranged (Bauer 1995a). For a concise, critical but not unsympathetic analysis see the article by Peter Marin (1982). Here is a brief chronology:

- Reich was born in 1897, served in the Austrian army in World War I, studied medicine, specialized in psychiatry, was an influential member of Freud's psychoanalytical movement in the 1920s. Quite early he decided that liberal sexual activity was necessary to sound human health.

- He moved to Germany and was active for a while in the Communist Party. Believing that "sex hygiene" was important as social and political policy, Reich founded "Sex-Pol," an institute for sexual politics.
- The Communist Party excommunicated Reich after a few years. He had also become too heterodox for Freud's psychoanalytic movement. Reich moved to Scandinavia, where for half-a-dozen years he practiced psychotherapy, founded a journal, and took up biological experimentation. He found apparently living "bions" in extracts from inorganic matter or dead plants and discovered the universal *orgone* energy.
- From 1939 in the United States, Reich was more preoccupied with his science than with therapeutic practice. He fell afoul of the law through selling "orgone accumulators" interstate and was sent to prison after refusing to offer any reasonable conventional defense.
- Even among those who revere Reich's memory and hold his ideas to be a still-unmatched advance, most agree that he was mentally ill during his last years (Edwards 1967, 1974); for example, he believed the Earth to be under attack by UFOs from outer space and was using his "cloudbuster" to destroy them.

  Some of Reich's supporters cite ample instances of the lack of acceptance and actual persecution he endured during his life that may have contributed to paranoia and megalomania. He died in prison in 1957.

Reich was highly educated, highly talented, a charismatic, multifaceted personality. He was far from unique in failing to practice as he preached; for instance, while he was clear about his own need to find sex wherever it took his fancy, he was not forgiving of those partners who cheated on *him* (Bauer 1995a; Higgins 1994). Here, in too compact a nutshell, are the Reichian assertions or beliefs that have been the most discussed:

- Neurosis is psycho*somatic,* not purely "in the mind." Character formation is accompanied by *armoring,* physical rigidity paralleling mental rigidity. The efficient removal of neurosis calls for body therapy in conjunction with talk therapy.
- Uninhibited orgasm is necessary for healthy mind and body. Adolescents must be given opportunities for satisfactory sexual release. Adults should freely engage in sexual intercourse. Social and political policy must respect this; the sexual frustration of the workers inhibits their political activity.
- Orgone energy is the universal form of energy; it is also the vital or vitalist force that is the essence of life. The sky's blue color comes from

orgone. Boiling certain inanimate materials produces cell-like structures that, upon absorbing orgone, become living cells, *bions.*

Orgone can be caught and concentrated in boxes consisting of layers of the right materials. Sitting in such an "accumulator" can cure a number of ills. Orgone energy can also be sprayed from these boxes by means of a funnel-shaped shooter.

There is also "deadly" orgone energy (DOR) that can poison atmospheres, climates, and regions.

Clouds can be made to produce rain by means of the "cloud-buster," an array of metal tubes grounded to some source of water, say, a stream.

Connectedness: What Makes a Pseudoscientist?

Reich's life may well be the type-specimen of the independent mind that fails to make effective connection with any intellectual community. He deviated too much from Freudian doctrine. He was unwilling to accept Communist orthodoxy. When his speculations about biology ran counter to mainstream views, he ignored even the very professionals he himself had asked for advice (Higgins 1994). Unwilling to negotiate with referees and reviewers, he founded his own journals and self-published his books. In his last years he ignored the offered help and advice of even his strongest supporters and closest friends and behaved suicidally in his interactions with the federal agencies and the law.

It seems natural to ponder how much in the progression to a sad end was owing inherently to Reich himself and how much to the manner in which his environment acted on him. Certainly he was strongly, even viciously attacked by those whose orthodoxy he flouted; but Reich seemed unwilling to make any practical compromises. An intriguing comparison is between Reich and Albert Szent-Györgyi (Bauer 1992a:61). Born within a few years of one another, one in the Austrian and the other in the Hungarian region of the Austro-Hungarian Empire, both were intelligent; talented; persistent, even head-strong personalities; charismatic; lifelong mavericks; internationally peripatetic. Both needed to and were able to support themselves and their work without holding regular academic positions. Both were true representatives of the scientific ethos of their time: self-consciously committed to a positivist scientific ideal and to the role of god-professors who paid only as much attention as they cared to, to others' opinions. Szent-Györgyi got a Nobel Prize; Reich is remembered chiefly as a crank. The only obvious difference is that Szent-Györgyi maintained crucial links with the mainstream. He submitted to the discipline of peer review, publishing in conventional journals, whereas Reich rejected criticism and founded his own periodicals.

Perhaps Reich's early activity in psychoanalysis and politics was poor intellectual preparation for science. Reich was in his late thirties when he took up scientific research and by that time may have been unable to subject himself to the strictures of established scientific knowledge. Perhaps in this he was like Velikovsky, who also tried to practice scholarship and theorize about science using the approach of psychoanalysis. But there the analyst's individual interpretations are paramount, whereas in science the researcher's work is subject to communal judgment.

The "mainstream" in some fields can vary drastically by geographic or political region. For Reich, support and opposition have both been cosmopolitan, as with some other schools of psychiatry or psychotherapy. Such as they are, Reich's connections to the mainstream have best survived the passage of time in psychotherapy and perhaps sociopolitical matters, not in biological or physical science.

Illustrative Points

Role of Leader — Reich's personal charisma seems to have misled some number of people into taking his "science" seriously. His outward behavior was not inconsistent with that of a mainstream scientific investigator. In the light of everyday common sense rather than of deep technical knowledge, his ideas could seem highly defensible. For those who lack familiarity with the real science of matters Reich dealt with, why would orgone be less believable than black holes, a bounded yet infinite universe, or "dark matter" (undetected yet supposedly responsible for 90 percent of all the mass in the universe)?

In the early 1980s I chanced upon the Wilhelm Reich Museum in Rangeley, Maine, and took its tour. The museum was advertised by flyers on local bulletin boards just as any other science museum might be. Having read Gardner (1957:chapter 21), I was bemused: the museum's Reich and Gardner's Reich could hardly be the same person.

Reich had worked in what is now the museum building. It struck me as a place in which my own Ph.D. mentor, a European of Reich's generation, would have been entirely comfortable. The apparatus on display, the photos of Reich in lab coat supervising workshops, the whole atmosphere seemed quite conventional. The tour guide was a teacher who had come upon the place some years earlier and had inquired about it of the locals; apparently a goodly part of the local population had accepted Reich as a serious investigator, a familiar figure as he drove around experimenting with his "cloud-busters."

As with Velikovsky, the mass of Reich's followers are unwilling to distinguish among the various points asserted in the guru's whole schema, in Reich's case his physical science, his biological science, his therapy, and his social and political opinions and practices. Many groups use "orgone" as part of their identification even when their primary concern is psychotherapy, which could after all stand quite independently of the existence or otherwise of orgone energy, although some therapists do advertise as Reichian with no mention of orgonomy (*The Whole Person*:50, 89). There is a Center for Orgonomic Studies, an Orgone Biophysical Laboratory, Orgonizado in Brazil, the Orgonomischer Informationsdienst in Germany, and so on. James DeMeo of the Orgone Biophysical Research Laboratory practices cloud busting and laboratory biology. The American College of Orgonomy publishes the *Journal of Orgonomy* and perpetuates Reich's latter-year megalomania: "Orgonomy has already demonstrated benefits in . . . diagnosis and treatment of emotional and physical diseases, prevention of armoring and neurosis; understanding of social problems; intervention to change disturbed work and family systems; influence of the weather; and comprehension of basic phenomena in nature from biology to astrophysics" (Crist 1995:1).

Schisms — As with Velikovsky, while the master was alive the doctrines and organizational arrangements were determined autocratically; with the guru's passing, schisms developed. There have been some overt disagreements between some of the various orgonomic organizations (Gardner 1988) over some mixture of intellectual differences and personal incompatibility (Kelley 1965). One organization publishes *Pulse of the Planet (Research Report and Journal of the Orgone Biophysical Research Laboratory)* while another boasts that *it* is "the only organization to publish an orgonomic journal of the high quality and depth of the *Journal of Orgonomy*" (Crist 1997:1). Alexander Lowen (1994) details where Reich went wrong and how his own "bioenergetics" accomplishes what Reich tried but failed to do. The Orgonomischer Informationsdienst (a Web site at <http://www.doit.de/orgon/editor.htm>) warns that not everything labeled "orgone" is thereby faithful to Reich.

"[L]itigation and threats of litigation . . . have characterized Reich's inheritors" (Kelley 1965:10). Mary Higgins, curator of the Reich Museum, has at various times sued or threatened to sue: Reich's widow (fourth wife) Aurora Karrer Reich; in the early 1960s, Charles Kelley who published an orgonomic journal for several years; around 1964, Paul Ritter, who practiced Reichian therapy; around 1968, Jerome Eden, and David Boadella, author of one of the biographies (Boadella 1974) of Wilhelm Reich, and Ellsworth Baker, founder of the American College of Orgonomy; around 1969, Peter Reich; around 1996, James DeMeo. For her part, around 1960 Aurora Reich sued Eva Reich, Wilhelm's daughter, for alleging that Aurora had stolen some documents. Eva Reich in turn sued Mary Higgins in about 1974. The American College of Orgonomy has at various times urged or forced dissidents out (for example, Trevor Constable, in about 1972, and Richard Blasband, around 1995) while others resigned in protest or disagreement (for example, Courtney Baker and others, around 1983, DeMeo and others circa 1995). Similar schisms have plagued orgonomic groups in other countries (Carlinsky 1998).

Support as a Matter of Fairness — As with Velikovsky, some prominent people befriended Reich: the psychiatrist R. D. Laing, the educator A. S. Neill, and the anthropologist Bronislaw Malinowski. As with Velikovsky, some were moved to support him primarily because of the way he was treated, in Reich's case by government agencies. The respected philosopher Paul Edwards wrote that "the virulence of the opposition to Reich only intensified my attachment, and for some years many of my friends and I regarded him as something akin to a messiah" (Edwards 1974:32).

In Velikovsky's case, a number of prominent supporters of the man overtly demurred from his science. In Reich's case, more people seem to have rather fully accepted much of his science. Thus even Paul Edwards, who distinguishes carefully between Reich's psychotherapy and his orgone science, goes on, "In fairness to Reich it should be added that a really unbiased investigation of his physical theories remains to be undertaken" (Edwards 1967:105). Still, "Reich's attractiveness was not the orgone theory . . . , which was received with a good deal of skepticism even by many of his warmest admirers" (Edwards 1974:32); Kathleen Tynan wrote, "What caught Ken's [Kenneth Tynan] interest was not so much the work on orgone energy as Reich's view that human nature is alterable and perfectible" (Tynan 1987:413–14).

Literature — Unless one understands something of the scientific issues or has come across critiques like Gardner's (1957:chapter 21), there are few clues that orgonomy is spurious rather than genuine science. The book-length biographies of Wilhelm Reich are all by sympathizers (Boadella 1974; Greenfield 1974; Mann and Hoffman 1980; Reich 1969; Sharaf 1983; Wilson 1981).

The Internet, as with Velikovsky, returns for "Wilhelm Reich" or "Orgonomy" mostly pro-Reich information, although a little browsing reveals schisms between different orgonomic groups and some general critiques of orgonomy. Skeptic's sites on the Web often have a section or links to debunkings of orgone energy.

Casual observers, nowadays as in the past, will not readily find clues to Reich's doubtful status as scientist; thus the *Chronicle of Higher Education* carried an advertisement "announcing **Lou Hochberg Awards**—topically focused upon **The Social Aspects of Wilhelm Reich's Discoveries—For Student Essays**

and Thesis/Dissertation Research." These are cash awards administered by the "OBRL" (1998a), the Orgone Biophysical Research Laboratory, of Ashland, Oregon, but "orgone" doesn't appear anywhere in the advertisement. Flyers from the laboratory (ORBL 1998b) describe it as "a non-profit science research and educational foundation, since 1978—building upon the discoveries of the late, great natural scientist Wilhelm Reich, M.D."

Recurrence — The most noteworthy recurrence of mistaken notions in Reich's work is his insistence that the "bions" produced from inanimate materials become living through absorption of orgone. Such "spontaneous generation" of animate from inanimate matter had been decisively disproven during the nineteenth century (Farley 1977), albeit others besides Reich had attempted to resuscitate it (Gardner 1957:chapter 10).

Reminiscent of polywater is the neo-Reichian notion of "energetic properties of water in its different structured states, a phenomenon believed to be intimately connected with the orgone charge of water" (DeMeo 1992).

Miscellaneous Points — As T. H. Huxley was "Darwin's bulldog" fighting creationism, and as Leroy Ellenberger is the indefatigable critic of Velikovskian and neo-Velikovskian catastrophism whom his former Velikovskian colleagues usually exclude from their journals and gatherings, so Joel Carlinsky (n.d., 1990, 1994, 1996) is the nemesis of orgonomists, who become apoplectic at the mention of his name. Because I had been in contact with Carlinsky I was sent very impolite letters and defamatory material about him.

Concerning rhetorical tactics, as with Velikovsky so it is with Reich. Hangers-on seek to portray brief encounters with Albert Einstein and other greats as scientific support. Thus we read that "Albert Einstein . . . invited Reich to his home to demonstrate his experiments and verified the basic phenomena Reich discovered" (Martin 1994); "Einstein confirmed . . . the temperature differential within the orgone accumulator" (DeMeo 1989:11). In reality "Einstein wrote to Reich that the temperature does indeed rise, but that there is a simpler explanation than concentrated orgone" (Gardner 1988:30).

Connections are asserted that are farfetched, or made with extremely doubtful bits of claimed science, as with the "independent or outright discovery of an orgone-like energetic principle . . . by scientists other than Reich, such as Giorgio Piccardi, Dayton Miller, Halton Arp, Hannes Alfven, Harold Burr, Louis Kervran, Frank Brown, Robert Becker, Björn Nordenström, Jacques Benveniste, and Rupert Sheldrake" (DeMeo 1994:16) as well as Paul Dirac and Thelma Moss (DeMeo 1989).

While a few of those mentioned have impeccable scientific credentials, the part of their work that might conceivably be related to orgone enjoys little or no credibility. Piccardi (1962) thought to have found simple chemical reactions affected by cosmic influences; Burr's (1972) work on "life fields" was never taken up by biology; the book of Thelma Moss (1974) would be classed by Arthur Clarke in the "Zeroth-Kind mystery" class; Kervran (1972) believes elements are biologically transmuted all the time in living beings; Nordenström, a professor of diagnostic radiology at the Karolinska Institute, claimed to have discovered a new electrical-biological circulatory system (Nordenström 1983).

Once more as with the Velikovsky affair, the assessing of Reich's work is not hampered by hoaxes or frauds. No one seems to have tried *faking* measurements of orgone energy or of temperatures in an orgone box (by contrast to making incompetently supported claims).

Summary

Reich's history—as of course that of many another maverick (Bauer and Huyghe 1999)—raises the question, How much independence of mind is *too* much? Surely Reich had excellent reason to reject the dogmatism of Freudian psychoanalysis and that of Marxist Communism. Perhaps one needs to stick closer to orthodoxy in science than in other fields; on the other hand, we know that a willingness to go against the conventional wisdom can be needed to make major scientific discoveries (Barber 1961).

As with Velikovsky, cool hindsight raises the question, Why were critics of Reich so apoplectic and vicious? He was a driven knowledge-seeker, not a deliberate charlatan, fraud, or con man. His propositions can hardly be equated with popular trash like newspaper horoscopes or healing magnets. Reich was intensely interested in matters of deep human concern, which is surely why he attracted some intelligent and competent people. His ideas about psycho*somatic* therapy continue to be regarded, for example, by Paul Edwards (1967, 1974), as a major contribution with significant implications for the mind-body problem. Once more, *odium scholasticum:* Reich infuriated the orthodox because he seemed, at least to much of the public and the media, to be a genuine scientist.

Loch Ness Monsters

Nessie, the Loch Ness monster, is the most famous topic in cryptozoology. I will cite here only the most pertinent points from my book-length study (Bauer 1986).

Connectedness

The search for Nessie is, so far as science is concerned, novel in facts, methods, and interpretation at the same time (Bauer 1983, 1986:152–53). The *methods* by and large are not methods accepted in biology to establish the existence of a hitherto unnamed species. Sonar, photography, eyewitness testimony, even myth and folklore may serve to provide acceptable information for known species, but not to establish the existence of a new species. So the purported *facts*, too, are unconnected with the biological canon. The *theory* is highly speculative, asserting a unique combination of reptilian appearance, mammalian temperature control, and respiration with only rarely observed visits to the surface; and demanding adaptation in less than about fifteen thousand years from saltwater to freshwater conditions and a population consisting of only a dozen or two individuals yet able to remain viable for that length of time.

The quest is carried on largely by nonbiologists, and there is little discussion of it in mainstream journals or at mainstream conferences. Nessie-hunting is separate from scientific investigation, both intellectually and in terms of the communities active in the investigation.

Illustrative Points

Rate of Progress — Nessie lore has not been as static as Velikovskiana or orgonomy; but any academic who had staked getting tenure on identifying the Loch Ness monster would have been long gone, generation upon generation (one "generation" with respect to tenure in academe being five or six years). Continuing reports of sightings have for decades added nothing of substantial value to the sightings reported decades ago; however there have been occasional films and, since the 1960s, a large number of sonar contacts. Yet there has been no meaningful progress in understanding. Theory has progressed from the 1930s suggestion of a single landlocked sea serpent to the 1957 scenario of sea creatures becoming landlocked about fifteen thousand years ago, after the last Ice Age; but the gain in plausibility thereby is hardly great.

Only one new book has appeared since the mid-1980s—a compendium, not any novel findings (Harrison 1999). The Internet is as fallible a guide as with Velikovsky or Reich.

Lack of Consensus — Nessie buffs differ among themselves over many points of detail as well as over the central issue of what Nessie is. Any given one of the "classic" photographs may be distrusted by some but given credence by others. Dinsdale's film is accepted by most but not all.

Many Nessie fans believe she is a population of creatures descended from the plesiosaurs. Roy Mackal favors a migrating primitive whale. Adrian Shine thinks it has to be a powerful fish, perhaps a sturgeon. Each suggestion has to select some of the eyewitness testimonies and ignore others; each suggestion encounters formidable objections. The only genuine consensus among almost all believers is that there is an unidentified species in Loch Ness. Most Nessie enthusiasts regard as unwelcome bedfellows those who suggest that Loch Ness monsters might be psychic entities (Beckjord 1988; Holiday 1986).

The debunkers are not unanimous either, except on the chief issue of agreeing that Nessies don't exist. For example, most of them accepted without question Christian Spurling's near-death-bed confession that he faked the most famous of all Nessie photos with a model of plastic wood mounted on a toy submarine (Anon. 1994a; Boyd and Martin 1994; Martin and Boyd 1999); but Steuart Campbell, one of the most incisive of critics and a longtime arguer against the reality of Nessies (Campbell 1991), finds that confession unacceptable (Campbell 1995) and continues to press his view that the photo is of an otter's tail. It's also worth noting that of the two investigators who interviewed and believe Spurling, one (Boyd) is a believer and the other (Martin) a disbeliever.

Arena of Dispute — The quest for Nessie and the arguments about it are played out in the public arena and not in scholarly venues. The general public maintains quite a strong interest in Nessie. Scottish tourism authorities believe that the monster attracts hundreds of thousands to the region each summer. Even though nothing substantive had been freshly discovered, during the late 1990s there were several television documentaries in which Nessie featured: in 1997 on cable channels A&E (Arts & Entertainment), TDC (Discovery), TLC (Learning); in 1998 on the History Channel's *In Search of History* series; in 1999 on public television (*NOVA* series) and TDC.

Miscellany — Debunkers lump disparate things as pseudoscience. As with cryptozoology in general, the debunkers' occasional epithet of "paranormal" is unjustified. Should a Nessie be found, biology will accommodate it as it did the coelacanth and the megamouth shark, or much earlier the platypus, oryx, and other unexpected creatures.

As to interdisciplinarity, which disciplines are relevant? Nessie hunting draws on eyewitness testimony (psychology, forensics), folklore, history, prehistory, remote sensing, sonar, photography—many fields besides biological science.

Eyewitness testimony plays a central, inescapable role. Nessie fans are fond of insisting that criminals would be readily convicted on evidence as overwhelming as the number and quality of Nessie sightings. There is, for example, that from Ian Cameron, a retired detective inspector, who was accompanied by another policeman who reported the same thing; *their* report was confirmed by seven people unknown to them watching from the other side of the lake. Even Adrian Shine, who pooh-poohs the notion of "monster" or prehistoric survival, finds himself unable to dismiss entirely the wealth of personal testimony he has had from trustworthy people.

There is an unexplained residuum. Nessie seekers regard as the *real* data all the sightings that haven't been shown to be misperceptions (seals, otters, waves) or hoaxes.

What is the evidence? Do the various types of evidence adduced by believers all relate to the same fundamental phenomenon, an animal? Or are eyewitness reports, folklore, sonar, and films and photos perhaps of altogether disparate types of events? What evidence there is cuts both ways (Bauer 1986:chapters 1, 2).

There is a lack of certified repositories for evidence. Evidence is often alleged to exist while not available for inspection. Some writers claim to have seen films of Nessies, albeit the films are locked in inaccessible bank vaults. Others claim to have seen Nessies described in an issue of the *Atlanta Constitution* dated somewhere in the 1890s, which is no substitute for giving the actual date and page number. Someone claimed a reference to "leviathans" in Loch Ness in a book of Daniel Defoe's, but again without specifying the edition or page. No one has been able to confirm those references.

There are grains of truth revealed through investigating the phenomenon. Nessie hunters have some points of success to display. Adrian Shine's efforts to describe the whole ecology and history of Loch Ness has already paid off in genuinely scientific knowledge—new species of worms and the use of loch-bottom cores to analyze changing conditions since the Ice Age.

Hoaxes and frauds have occurred with great frequency. In addition, commercial exploitation is at a high level. Scotland's economy, a fortiori that of the Highlands, depends heavily on tourism. Even the very bad 1996 Hollywood movie *Loch Ness,* starring Ted Danson, Joely Richardson, and Ian Holm, is said to have boosted tourism markedly. The exhibition at the Drumnadrochit Hotel made a fortune for its proprietor. Every type of souvenir that can bear a Nessie logo, does.

As with Velikovskiana, the appeal of Nessie-questing seems to be primarily intellectual, the same sort of curiosity that drives scientific investigation.

Shibboleths abound, about Nessie sightings and runs of salmon, the creatures' sensitivity to noise, and more (Bauer 1986).

Debunkers say, "Impossible!" They claim, for example, that Loch Ness doesn't contain enough food for even a small population of monsters of the size claimed by Nessie fans (Cambridge Loch Ness Expedition 1962; Tippett 1983); or that no such large creature could have survived as a large enough population to remain genetically viable; or, to those who say Nessies are evolved plesiosaurs, that this is impossible in such cold water for creatures supposed to lack temperature regulation; and so on.

Closing ranks, enthusiasts are unwilling to criticize one another in public (Bauer 1986:82–83).

Summary

Again as with Velikovsky and Reich, it's hard to see why the search for Nessie should be equated with popular trash like newspaper horoscopes or healing magnets. Most Nessie buffs are knowledge seekers, not conscious charlatans or confidence tricksters.

Extrasensory Perception

Belief in nonmaterial forces, energies, or spirits stretches back into prehistory. Scientific or quasi-scientific studies began in the nineteenth century. Nowadays such investigations are generally called parapsychology and the underlying force or phenomenon or human ability is often called *psi*.

The literature is vast. General histories of the field have been written (Beloff 1993; Broughton 1992) as well as historical-sociological analyses (Mauskopf and McVaugh 1980; McClenon 1984) and debunking works (Alcock 1981, 1990; Kurtz 1985). Here will be considered only how the comparisons and generalizations about anomalistics made in earlier chapters apply to parapsychology; and how the viewpoint taken in this book differs from that of a typical disbeliever's critique.

Connectedness

It is the *theories* of parapsychology that separate it most sharply from mainstream science, for parapsychology hinges on the existence of human information- or energy-transmitting abilities that are not mediated by tangible physical mechanisms.

The raw *facts* of parapsychology are not in dispute; it is their interpretation that is problematic. A spontaneous instance of apparent mental telepathy or precognition or clairvoyance is treated by believers as a manifestation of *psi* while to disbelievers it is a coincidence or mistake or hoax. Controlled studies that have led to results apparently different from what could be achieved randomly or by chance are interpreted by believers as measures of *psi* but by disbelievers as experiments improperly controlled and open to statistical fallacies if not to outright fraud. In any case, the data are not generally found in mainstream texts or periodicals.

Though mainstream science rejects the idea of *psi*-like interactions, such mainstream fields as philosophy and psychology acknowledge that the interactions between mind and body are a mystery, the "mind-body problem": ideas and emotions in every individual human mind are somehow coupled to chemical and electrical events in that individual human brain. *Psi* would be the same *sort* of interaction but not limited to the inside of each individual animal body.

The *methods* of parapsychology differ little from those used in some specialties in psychology. In biofeedback studies as in psychokinetic studies, subjects use their will or wish in the attempt thereby to influence physical variables—in the first case inside their own bodies, in the other, outside—but much the same experimental conditions apply. Mainstream psychology as well as parapsychology makes use of anecdotes, for example in clinical work and psychotherapy. Statistical analysis of results is quite common in psychology, though it may be somewhat more central in parapsychology.

So parapsychology connects closely with psychology in method, but differs sharply in theoretical stance. Parapsychology is therefore less disconnected substantively from the mainstream than is Velikovskiana or Nessie-dom. That parapsychology is intellectually connected to the mainstream in its methods (in other words in one of the three aspects of facts, methods, and theory) makes it rather an instance of "premature science"—though of course not every instance of premature science eventually becomes mature.

The parapsychological community is largely separate from that of psychology though there is some connection to the mainstream. The Parapsychological Association is affiliated with the American Association for the Advancement of Science; there is a chair in parapsychology at Edinburgh University, a research fellowship at Cambridge (Epstein 1993), laboratories at a number of universities. Many eminent scientists have devoted some effort to parapsychological work, recently as well as in the late nineteenth century. A few prominent psychologists have written favorably about parapsychology or have done some work in it. William McDougall was eminent in psychology and also founder of the Parapsychology Laboratory at Duke University. Psychic research, like alternative medicine, has been quite palatable to large sections of mainstream society in Britain and parts of continental Europe.

There seems little justification in all this for dismissing parapsychology holus-bolus as rank pseudoscience, as many debunkers do.

Typical Criticisms of Parapsychology

James Alcock professes psychology at York University (Canada), is a member of the executive council of CSICOP, and has written some substantial critiques of parapsychology (Alcock 1981, 1990). His article (1996) in the *Encyclopedia of the Paranormal* will be used here as an exemplar of scholarly criticism of parapsychology.

Such scholarly criticisms of parapsychology raise a number of telling substantive points. Certain implications of *psi* would seem to contradict much sound, established empirical knowledge:

- No one has ever won lottery after lottery, not even a couple or three; obviously even the most gifted psychic cannot deliberately see even a little way into the future.
- No one has claimed Randi's offer, now at over a million dollars, for a demonstration of *psi* (James Randi Educational Foundation 1997).
- No one has successfully performed psychic cures of person after person.
- Physical constants give the same measured results within extraordinarily close limits, for a variety of experimenters, even though some of those surely wanted desperately to find something different, since only a new value would bring them any kudos.

So the power of *psi,* if it exists, might seem restricted either to marginal aspects of reality where physical consequences are tendencies rather than strictly determined (Jahn and Dunne 1987) or to certain special occasions, say, deathbed visions—in essence, miracles.

Alcock makes these and other eminently sound points: the disconnectedness of parapsychology from the mainstream, the lack of replicable experiments, and the difficulty of interpreting small statistical effects. But none or all of those entails that *psi* does not exist or that people should be ridiculed for trying to pin it down. Moreover Alcock lapses into invalid assertions, for example:

Alcock (1996:250): "*Unfalsifiability.* In regular science, hypotheses are not considered useful unless there is some way of disproving them."

Comment: As already mentioned, this was never a criterion in science. It was a suggestion by the philosopher of science Karl Popper, never accepted by his colleagues and later abandoned by Popper himself after he got into trouble through application of the notion to, for instance, the theory of evolution. This Popperian criterion has also been (mis)used to class psychoanalysis as pseudoscience, which ought to make psychologists particularly hesitant to invoke it.

Alcock (1996:242): "The popularity of belief in extrasensory perception . . . [is] generally . . . at a peak during periods of great societal change or crisis."

Comment: Was Britain in the late nineteenth century, when the Society for Psychical Research had its heyday, experiencing great change or crisis? The Victorian era is more usually described as a notably stable social epoch. Perhaps religion was undergoing something of a social crisis around that time as a result of Darwin's theory of natural selection. But the majority of people continue to maintain belief in religion as well as in extrasensory perception, which leaves Alcock's asserted correlation less than compelling.

Moreover, Alcock is simply mistaken in citing in this context revolutionary discoveries in science, like X rays and radioactivity, as examples of intellectual crisis. Those postdated by a good decade or more the founding of the Society for Psychical Research, which came (in 1882) during the still self-satisfied Age of Science (Knight 1986).

This generalization of Alcock's qualifies less as a substantive point than as a debunkers' shibboleth: *because* extrasensory perception is in their view pseudoscience, the reason for its prominence in the public mind has to be psychological in origin, and so one postulates social-psychological crises to explain psychic belief.

Illustrative Points

Rate of Progress — Science progresses but pseudoscience does not. But it is debatable whether or not *social* science progresses as natural science does. Is our understanding of human psychology markedly superior to that of Greek playwrights twenty-five hundred years ago? There are certainly more methods now of obtaining information about human psychological behavior, in some cases even semiquantitatively repeatable statistical information; still there are no indubitably repeatable experiments in human psychology.

Parapsychology seems comparable to psychology in this. It has seen improvement in method, changes of focus, but no unequivocal progress in understanding. In the nineteenth century, investigations were primarily of physical mediums, who are nowadays regarded by most people—including most believers in *psi*—as having been frauds. Modern parapsychology began with J. B. Rhine's laboratory studies of telepathy and clairvoyance ("card guessing") and psychokinesis (Mauskopf and McVaugh 1980). Rhine's use of "Zener cards" with five clearly distinguishable symbols was a notable advance in method and produced semiquantitatively repeatable statistical results. The introduction several decades later of random electronic generators and computerized record-keeping and analysis represents very significant experimental progress, including apparently foolproof safeguards against tampering and fraud. "Ganzfeld" studies mark another important innovation, the masking of sensory signals as far as possible in order to enhance the *psi* signal; indeed one longtime prominent critic of parapsychology declared himself convinced that cumulative analysis of Ganzfeld data gives clear evidence of extrasensory perception (Bem and Honorton 1994).

Compared with psychology, everyday "normal" parapsychology is limited by the small number of people who have the resources to take advantage of the modern advances in experimentation. The lack of replication or building upon apparently significant results is striking. For example, studies of telepathy while dreaming gave positive results (Ullman et al. 1974) and it would seem natural to continue such work, varying conditions to pin down something of the characteristics of *psi;* yet it seems that only one attempt to reproduce the work was undertaken, unsuccessfully (Alcock 1996:247), and so the original work remains as a solitary datum even though many parapsychologists continue to cite it as classic. Again, there is wide agreement that the Princeton Engineering Anomalies Research (PEAR) effort has evolved apparently foolproof methodology that appears to deliver statistically significant evidence for some anomalous interaction between consciousness and the environment, albeit the magnitude of the effect is very small. How invaluable it would be, for that group of researchers above all, if there were, somewhere else in the world, another laboratory seeking to do the same sort of thing. Replication of the results would be a striking and most welcome confirmation, whereas failure could lead to further refinement of apparatus or analytical approach. In the absence of another such group, the PEAR investigators must be left with the nagging thought that there might somehow, somewhere, be something they could have overlooked.

The Evidence Cuts Both Ways — On a number of points parapsychology illustrates how believers and disbelievers can offer plausible yet opposite explanations.

1. Shyness effect (that psychic events are observable only when the experimenters are believers):

   Of course, according to the true believers, that's precisely what one would expect of a human ability that isn't easy to exercise; anything antagonistic or untoward in the situation may suppress it.

   Not at all, say the debunkers, what we have here is involuntary self-deception by the believing investigators, so sure of the effect that they fool themselves.

2. A supportive environment is necessary for psychic events to be observed:

   Very much as with the shyness effect—of course a psychic person will not be able to deliver peak performance in an unsupportive atmosphere; everyone knows about the "home advantage" in athletic contests, the spur to performance of a cheering crowd and the depressing effect of a quiet or unfriendly one.

   The debunkers, of course, recognize that a supportive environment means an easier opportunity for the claimed psychic to cheat, being less rigorously observed and controlled.

3. Decline effect (that any given person's psychic performance decreases in successive trials):

   Just what you expect of a human ability, say the believers; we all get tired.

   No, according to the debunkers, this is just statistical regression to the mean, the initial above-chance results were just the upper end of the bell curve (more on this later).

4. No repeatable experiments:

   Believers: some human faculties are spontaneous, involuntary, albeit capable of being harnessed, for example we have learned via bio-feedback to control physiological functions like pulse rate; it's just that no one has yet learned to control *psi.*

   Disbelievers: the anecdotes about premonitions, hunches, and the like are just that—anecdotes, not evidence.

5. Above-chance statistics: those anecdotes, and many trials even though not exactly repeatable, add up to enormous odds that something beyond chance is at work.

   Believers: that something is *psi.*

   Disbelievers: many things can explain an apparently significant away-from-chance result—use of an inappropriate statistical test, inappropriate controls, unrecognized systematic or singular errors, subconscious or deliberate fraud. Any of those is more likely than *psi.* Moreover, even if all "normal" causes have been eliminated, "to interpret an extra-chance scoring rate as evidence of some paranormal power is no more justified than taking it to reflect the whimsical divine intervention by Zeus or some other deity" (Alcock 1996:245).

6. Varieties of psychic phenomena:

   Believers: Clairvoyance, telepathy, precognition, psychokinesis, psychometry, dowsing, out-of-body experiences, reincarnation. All those and more, like auras, are a reflection of the fact that humans are not merely material objects, they have a spiritual or psychic essence that can transcend the limitations of the material body.

Debunkers: What evidence is there that those things are related to one another? Isn't it possible that there could be telepathy, minds communicating with one another, yet no precognition or clairvoyance? After all, our senses of sight and of hearing and of touch are separate and independent of one another, mediated by different mechanisms for turning the external signal into nerve impulses. Thus one might explain mind-to-mind telepathy by analogy with radio transmission at "other frequencies" but environment-to-mind transmission of information by some other mechanism; and neither one entails the existence of the other.

Shibboleths — Parapsychology brims over with shibboleths that conduce to the interpretation of *psi* as a human capability: the just-mentioned decline effect and the need for a supportive ambience and believing investigators. That *psi* displays such apparently human characteristics is most welcome, for in other respects *psi* behaves like no other human ability or sense. Thus its efficacy is not influenced by distance nor by intervening objects whereas our other senses are, and most remarkably we do not get better at using *psi* as we practice and try, unlike the improvement we can attain with practice in physical or intellectual activities. So the decline effect is a very congenial shibboleth for believers and widely acknowledged. However, it *is* a shibboleth and not a fact. Perhaps the most carefully controlled studies ever performed, at the Princeton Engineering Anomalies Research laboratory, failed to reveal a decline effect (Jahn 1982; Jahn and Dunne 1987).

If the "decline effect" is spurious, one can suggest how it might have come to be accepted. If in a number of instances "psychic performance" was actually only the small proportion of all attempts that, by chance, yielded a high rate of success, then those to whom that happens will not continue to have consistently high success. Test a thousand potential clairvoyants, and some will do better than average and others worse. Say about fifty of them perform significantly better than chance would predict. Now take those fifty and test them again, and a couple will again do that well, but the others will have shown a decline effect. Take the two or three still successful performers, test them again, and they will likely do no better than chance—they too have now demonstrated the decline effect.

In other words, on a bell curve some performances are always out on a tail; but when those performers are subjected to another trial, the result is a new bell curve on which again only a small proportion of all performers are out on a tail. It is not unlike the stock-market tipster mentioned earlier: in each successive trial, the number of people is smaller and the number of successes is smaller though their proportion to the whole is the same; each individual has a different record of successes, but all of them eventually experience failure. Statistical parlance describes this as regression to the mean. But believers can interpret such sequences as a decline effect through focusing attention retroactively only on those individuals who had at first appeared to be gifted.

Recurrence — For unorthodoxies to recur, they must first drop out of sight. There seems to have been no time when people did not have everyday experiences of déjà vu, premonitions, clairvoyance, and the like. In the past, organized religion at times suppressed or oppressed such matters as sorcery or witchcraft.

As far as focused scientific investigation is concerned, that seems to have begun only in the nineteenth century. One instance of a recurrence, in anomalistics if not in parapsychology per se, may be the "channelers" of the contemporary UFO scene, who seem very like the spiritualist mediums of the late nineteenth and early twentieth centuries.

Consensus — In psychology, there is notable lack of consensus about salient matters, especially in theory; it stretches from reductionist behaviorism to mystical "humanistic" psychologies.

In parapsychology, many but not all believers presume that the various claimed psychic anomalies belong together as instances of *psi;* but nonbelievers can not only interpret any or all of the claimed events as mistakes or deception, they can also deny that telepathy, clairvoyance, precognition, out-of-body experiences, reincarnation, and the like have anything indubitably in common even if they are real phenomena.

There is no consensus among parapsychologists over what the best evidence is for the reality of *psi.* There is no agreement over the validity of "laws" that have occasionally been proposed in parapsychology, like the decline effect or the influence on experimental results of the experimenter's belief. There is no agreement over which of the various claimed manifestations of *psi* are genuine.

Arena of Dispute — Belief in most of these various manifestations of *psi* is the majority view among the general population, at perhaps about 80 percent in North America (Alcock 1996), but a minority view among the "experts"—mainstream psychologists and pundits who purport to speak for "science."

Fraud and Hoax — Many psychic charlatans have been exposed. Perhaps one of the most startling frauds in parapsychology was the mathematician S. G. Soal. The frequency of such hoaxes means that faking must be guarded against in parapsychology as a perennial possibility whereas mainstream psychology as well as science generally can still regard fraud as unlikely in the extreme. Not that it fails to occur; seminal studies on treatment of hyperactive children were falsified (Byrne 1988; Zurer 1988) and there is a widespread belief that Cyril Burt faked essential data in his influential studies of twins (Mackintosh 1995).

Summary

Though parapsychology may not seem scientifically impressive when compared to an idealized view of "science," a cool analysis of its methods, facts, and theories marks it as comparable to instances of "premature" science rather than of pseudoscience. In a number of respects it is quite comparable to psychology itself; and both fields came into existence as scientific or quasi-scientific endeavors in the late nineteenth century.

Serious Anomalistics

Parapsychology, cryptozoology, and ufology are the "Big Three" serious anomalies, but they are not the only ones. In all of them I believe a disinterested observer will find, as in those briefly attended to above, rather little connection with mainstream thought and mainstream organizations. In each case, however, one finds *some* connections, intellectual or social or both. In each case one finds a few accomplished mainstream scientists or scholars who, for one reason or another and despite their peers' disapproval, refuse to dismiss the matter out of hand.

There is nothing in the practice of the serious investigators that could be labeled charlatan-like in the sense of deliberately seeking to hoodwink others for personal profit. Moreover, searches for Nessie or *psi* are not obviously irrational. Nevertheless, in each case one does find pundits who cheerfully decry the studies as pseudoscience, lumping them together with blatant foolishness like pyramid power or blatant commercial exploitation like the magnetic healing alluded to earlier. Such lumping and such denigrating are not grounded in any criteria accepted as valid in philosophy of science nor are they grounded in empirical observations of how real science works.

7 Deciding about Scientific Anomalies

The chief conclusion from both the generalizations in earlier chapters and the cases in chapters 4–6 is this: the practices of anomalistics, natural science, and social science are all determined by the nature of the subject being investigated. Parapsychology and psychology differ from one another rather as premature science differs from relatively mature science. Seeking Bigfoot differs from field studies of the gorilla in a number of ways that stem basically from how little is known about the former compared with the latter; but in other respects there are considerable similarities, such as reliance on reports from human observers. Searching for Bigfoot is not pseudoscience *because* it isn't done scientifically or *because* it's carried on by nonscientists including some incompetent people. It is just that the quest itself is inherently and inevitably a quixotic adventure that must bet on success against all the odds. It would still be much the same sort of pursuit even if it were joined by task forces from the National Academy of Sciences and the Corps of Engineers. Studies of *psi* remain anomalistic studies even though sponsored by the Defense Department.

Anomalistics differs from the natural and social sciences primarily in that the latter are well-established disciplines while anomalistics is not. From that flows the differences pointed to in the foregoing chapters: the relatively disorganized state of research and literature and groups studying anomalous phenomena and the concomitant lack of intellectual guidance for making judgments about research proposals and research reports. But the questions asked by serious anomalistics—say, whether psychic abilities exist—are not a priori different in intellectual character or quality from the questions asked in natural or social science. Indeed some of the claims are

> of the very highest quality—some of the best that has ever been produced by the minds of men (and, dare I say it, perhaps even more frequently by the minds of

> women—who seem to provide the majority of guides to the Spirit World). Anyone studying . . . [these claims] can hardly fail to be entertained, amused, sometimes saddened, and always instructed. It teaches a great deal about human psychology and the motives controlling the behaviour of even the most intelligent and rational members of that peculiar species, *H. sapiens.* (Clarke 1984a:4)

Seeking new knowledge is fraught with pitfalls. Being inside a scientific community affords some protection against them in the normal course of events. But when it comes to potentially revolutionary discoveries, there's no foolproof shortcut by which to distinguish what the future will consign to pseudoscience from what it will nominate for a Nobel Prize. Members of the scientific community whom we now credit by hindsight for the great breakthroughs had, in their time, to go against the grain of their colleagues' staid conventional wisdom just as those did whom we now recall as cranks. As I. J. Good (1998) has remarked, geniuses are cranks who happen to be right; as equally, of course, cranks may be geniuses who happen to be wrong.

For every generalization that might guide our judgment in separating wheat from chaff, science from pseudoscience, there are exceptions to be made; especially, of course, in the hard cases that are also the ones most likely either to yield breakthroughs or to turn out to be spurious.

The absence of sharp boundaries and the lack of definitions or formulas for readily and quickly distinguishing true from false, real from spurious, does not however doom us to complete uncertainty. Nor does the lack of certain answers mean that we run great danger through some ever-present possibility of lapsing into false beliefs. It is possible to think rationally about things like UFOs or psychic phenomena without either accepting their reality or rejecting it.

The following suggestions seek to merge the generalizations offered in chapters 1–3 with empirical observations like those in chapters 4–6 to indicate how one might go about confronting a startling knowledge claim. At the outset it makes sense to distinguish between commercial trash and serious anomalies. A further vital distinction is between beliefs and actions. Assessing the evidence poses great challenges, but the quest can be rewarding.

Demarcation: Substance or Trash?

Topics often lumped together under "pseudoscience" bear no obvious relation to one another, in substantive content (psychic events, cryptic animals, pyramid energy) or in the degree to which that content is absurd in the light of well-tested current knowledge. These subjects can however be separated, without leaving too large a fuzzy borderland between, into two categories: commercial huckstering on the one side and fascination with genuine mysteries on the other.

To Investigate or to Buy

It's not so difficult to discriminate suggestions that something is worth looking into from assertions that something is already established. A biochemist friend recommends that I take Co-enzyme Q-10, tells me why he takes it himself, explains why it might help and is very unlikely to harm, and invites me to explore the issue further for myself. That is one thing. The glossy *Journal of Longevity: Medical Research in Preventive Medicine Fields* I receive unsolicited is a very different matter. It assures me of wisdom about health and longevity that only Dr. Kugler and his associates command, not the rest of the medical profession. I am invited not so much to read and inquire further but to purchase at a cost of only about a dollar per day a host of special formulas: memory caps that "protect and build brain function," omega blend "immune and cardio enhancer," and so on. I have good reason to believe that my biochemist friend is seeking to share knowledge that stands some reasonable chance of being sound; I have equally good reason to believe that the *Journal of Longevity* cares primarily to get my money.

It would be as foolish at once to trust the *Journal of Longevity* as it is foolish to sign up for cheaper phone service or new credit cards or anything else on the basis of a solicitation by telephone, or to accept the generous offer made at the front door to repave the driveway at very low cost with materials remaining from another job and therefore available at extraordinary savings, or to sign up for pest extermination after an unsolicited free inspection.

Of course it's possible that I could get a good driveway sealing done at some savings; perhaps I could save something on my telephone bills; perhaps Dr. Kugler is right about the benefits of some substances that are not part of the conventional arsenal of drugs or dietary supplements. I shouldn't necessarily disbelieve the claims made to me by people who seem to be primarily after my money; but surely I should hesitate before taking the action of paying them.

That something has a commercial aspect to it does not automatically mean that it's spurious and intended only to exploit. Commercial conflicts of interest are inherent in a free-market society; they are suffered by doctors, hospitals, and HMOs, by pharmaceutical companies, health spas, and by researchers who need research funds and facilities. Serious investigators also have to earn a living, and if they don't have private means or an occupation that allows them the necessary free time or resources, then they have to combine anomalistic investigation with raising money for it. That conflict of interest is ever-present for anomalists as it is for scientists (Bauer 1992a:92–98) as well as for debunkers; note, for example, the continuous fund-raising efforts of CSICOP and its affiliates (e.g., Center for Inquiry 1996) or Randi (James Randi Educational Foundation 1997).

So automatic dismissal of apparently exploitative anomalist claims may not be justified, since not only pseudoscientific charlatans are on the make. Still, when money seems to be the prime objective, we should be doubly on guard. That we still say "caveat emptor" indicates that to have been true for a couple of millennia at least.

How Could That Be Known?

Faced with an unlikely claim, it's worth asking, "How could that have come to be known?"

Consider the range of dietary supplements, herbs, vitamins and the like whose benefits are touted in mail-order catalogs and pamphlets. A brochure from Gero Vita Laboratories in Toronto, Canada asserts that age spots on the skin "are known as lipofuscin. In the brain, lipofuscin forms a brown slime on the delicate neurons. . . . A deficiency of some very important nutrients has been correlated with substantial increases of lipofuscin. Those nutrients are RRR-a-tocopheryl, selenomethionine, glutathione, chromium and dimethylaminoethanol (DMAE)." In addition we need magnesium lest our brains tire, and the magnesium needs pyridoxine to work properly; and "Men should be especially concerned about maintaining high zinc levels because it is the most important nutrient for sex drive." "Another antioxidant critically needed by the brain is ascorbyl palmitate"; and *Hypericum* helps against depression; and we also need something to counteract the monoamine oxidase whose production in the body starts to climb after age thirty-five. So: take "GH3" ($109.95 for a six-month supply, reduced from the regular price of $239.95) and enjoy the benefits others have experienced—reduced days of sickness; joint pain and gray hair gone; loose skin tightened; protection against heart disorders, tumor, and cataract; relieved asthma; digestion improved; not tired any more; pain is gone; blood pressure normalizes; memory improved; skin softer; hair grew fuller; stopped desire for booze; helps diabetic; skin doesn't bruise now.

Pause to contemplate how many clinical trials would be required to test each one of those claims. With about a dozen ingredients, it's not unlikely that there will be interactions, so possible synergy between each pair of ingredients must be tested; and between the ingredients and common substances in food as well as commonly used medications like analgesics and antihistamines. To establish Gero Vita's claims, then, would call for hundreds of trials, and one can reasonably doubt that they were ever carried out.

This line of thought indicates at once that unqualified claims à la Gero Vita of a number of benefits from a combination of "necessary" substances are most likely not based on sound tests. At best they are guesswork extrapolating current understanding of various biological mechanisms together with anecdotes that fail to allow for placebo effects.

For example, my friend's recommendation of Co-enzyme Q-10 is based on the known action of that substance in the energy-generating reactions inside cells and the known fact that aging is accompanied by a decline in the production of substances needed in those reactions. If more Q-10 could be supplied in the right places then it would quite likely have a beneficial effect; but it remains unknown whether taking it orally delivers it to the places where it's needed. On the other hand, it's unlikely in the extreme that oral ingestion of this particular substance (a small molecule of simple structure) could be harmful in any way, so there's probably no danger in using it.

As to measuring benefits, that's extremely difficult. Innumerable factors influence how we feel, and to discern the effect of a single one while all the others are changing as well calls for lengthy and careful observation. Thus I might take Q-10 in alternate weeks and keep careful records of my relative energy levels; and still there would be quite a few confounding factors. My feeling of being energetic or otherwise varies seasonally. Such events as the birth or death of a relative would override for some period of time other factors in "feeling energetic." And so on; when it comes to attempting to assess benefits whose only measure is how we feel, any conclusion can be tentative at best. So long as it doesn't do any damage, and I don't feel the worse for it, I can give Q-10 and my friend's intuition the benefit of the doubt; and I can easily afford the fifty cents or dollar a day that it costs. If I seem to note enhanced energy and well-being, that's not proof that the Q-10 is responsible. It's just an interesting story, one I can feel comfortable with in my personal affairs, but it doesn't mean that others would have the same experience. Anecdotes are rightly treated by science as doubtful evidence at best.

So it's most unlikely that there is sound, substantive experience underlying the claims made in Gero Vita's brochure for GH3, or in any similar offer. If it's not going to cause damage, and if you can afford the price, there may be no strong reason not to take it. But it is not always clear that ingestion of sometimes large amounts, even of substances that are naturally present in the body, might not cause harm. For example, the fat-soluble vitamins (A, D, E, F, K) accumulate in the body, and too much of them can be harmful; continued eating of polar-bear liver, 30 grams of which contain 450,000 units of vitamin A, has caused peeling of the skin from head to foot (Jones et al. 1987:554).

In considering whether or not to put out money for GH3, one may also be guided by clues separate from any of the technical issues just noted. The flyer for Gero Vita carries an advertisement for the *Journal of Longevity,* and the latter carries an advertisement for other items available from Gero Vita International. This offers seventeen medicaments: Memory Caps; Moderex to reduce craving for alcohol; Nic-Away to stop smoking; Omega Blend immune and cardio enhancer; and so on. In *all* seventeen cases, the "reg" price is between $39.95 and $44.95 for a month's supply, but available at a special reduction for $29.95 (or six months for $109.95). Do all the disparate ingredients of these different medications really cost the same? Or could it be that the price is set only according to what the market can bear, irrespective of the cost or value of the ingredients?

When a theory makes a prediction that is fulfilled, that gives us further confidence in the theory. The preceding paragraphs were based on material I had received up to 1998. In March 1999, Gero Vita advertised a new product, Sexativa Plus, a miraculous Chinese discovery safer than Viagra but performing the same function. Before I'd read to the end of the pamphlet, I made a prediction: a six-month supply would cost $109.95, reduced from the regular $239.70. Sure enough.

Unorthodoxy in General

Unorthodoxy has certain implications, whether the unorthodoxy be in intellectual questions as with anomalistics or in religious, political, or social matters. So anomalistics has certain similarities with other types of unorthodoxy; and in some cases there are not merely similarities but actual overlaps with what are normally regarded as religious or political issues, regarding creationism, say, or political assassinations.

Belonging to a Minority

Is it foolish or even dangerous to take anomalies seriously? Is it foolish or even dangerous to be unorthodox? To be in a minority? Is it foolish or dangerous to be in the company of people like Linus Pauling, Martin Fleischmann, or Wilhelm Reich?

Surely not always or in every respect. "All progress has resulted from people who took unpopular positions," said Adlai Stevenson (1954). "All progress depends on the unreasonable man," said George Bernard Shaw (1946:282). "Ah! don't say you agree with me. When people agree with me I always feel that I must be wrong," said Oscar Wilde (1901:202).

Being in a minority, taking unorthodox positions, may ultimately prove to have been the right thing. In the meantime, however, it can be uncomfortable.

One who holds a minority view over *anything* experiences as a result certain types of interactions with other people and with social institutions that those who stick to the conventional wisdom do not; those typical interactions can be other than pleasant.

On the other hand, being in a minority can also offer satisfactions or benefits. Thus the Christian Coalition "can fend off mainstream criticism by assuming the posture of the persecuted"; its guiding light, Ralph Reed, "had watched as blacks, women, homosexuals and the disabled parlayed victim status into political recognition, and he saw evangelicals as a similarly persecuted minority" (Greenberg 1998). The role of underdog, of the David who potentially may slay Goliath, can be attractive; so in intellectual anomalistics, the mavericks quite often come to compare themselves with Galileo.

That comparison encompasses not only the minority status but also that of subsequent victor, which captures the typical ambivalence that the unorthodox tend to have. On the one hand they want the mainstream to join them; on the other they derive a certain psychic satisfaction from their presently underdog situation.

Similarities among Unorthodoxies

Among the similarities between scientific and nonscientific unorthodoxies are these:

- *The border between mainstream and unorthodoxy is neither stable nor easily defined.* Often some fuzzy areas or topics defy classification; and some occasionally move from mainstream to unorthodoxy or vice versa.

  Scientific: N rays, polywater, and cold fusion began entirely in the mainstream but are, by hindsight, often labeled as pseudoscience. Acupuncture was labeled pseudoscience until a couple of decades ago but nowadays is quite respectable in many circles. Such fields as bioelectromagnetics straddle the border.

  Christian Science is often labeled as pseudoscience, yet it is recognized by mainstream society as a legitimate religion. Scientology too is denigrated as pseudoscience by groups like CSICOP but has won legal recognition as a proper religion—in some countries, that is to say, though not in others.

  Political: At various times and in various places, almost every sort of political view has been sometimes mainstream and sometimes unorthodox: Communists, socialists, and libertarians have been unorthodox minorities in most of the world in the nineteenth century and in some of it in the twentieth, but still Communism has been and still is orthodoxy in significant sections of the globe.

Religious: Like Christian Science or Scientology, the Mormon Church or the Unification Church ("Moonies") and others as well are sometimes accepted as mainstream and at other times or in other places rejected as "cults." Apocalyptism "in its purest manifestations 'is the refuge of an embattled minority, [but] it is not always sharply distinguished from what we regard as the mainstream of Western thought'" (Thompson 1997).

Social: The arts traditionally occupy something of a borderland, some aspects of which are accepted and even prized in the mainstream while other aspects reach beyond the pale. Arguments over funding of the National Endowment for the Arts often focus on what is art and what isn't. As an instance of what some find utterly repugnant but others accept as art, the 1990s offered certain public performances: "Pain is the motivation for S&M artists, who fascinate and disgust audiences by mutilating themselves on-stage" (Hamilton 1997:11).

Distinctions between "highbrow" and "lowbrow" in art, literature, and music are anything but fixed, as "camp" or "pop" art go in and out of elite fashion. Even within a given genre there may be such divisions, for example between "authentic bluegrass" and the "commercialized mainstream music establishment" (Smith 1998).

Sexual: For long periods in much of Western culture, anything but heterosexual relations and monogamous marriage was beyond-the-pale unorthodoxy. In some parts of the world, however, polygamy is conventional.

Nowadays in Anglo cultures, homosexuality and bisexuality have moved from beyond the pale to largely acceptable; at the same time in many other cultures, homosexuality remains beyond the pale.

To avant-garde American academe, lesbian and gay sadomasochism is a valid life-style (De Russy 1998; Kimball 1998), but not to the wider society (Anon. 1998b).

Altogether, what is labeled as "deviant" sexuality varies from time to time and from place to place (Bryant 1982).

- Consonant with the fuzziness of the border between mainstream and orthodoxy, *some distinctions are purely semantic,* based on names and not on substantive characteristics of the things named. Using the right name gives legitimacy to otherwise frowned-upon objects or practices.

  Scientific: Some examples have already been given: "mesmerism" is pseudo but the same phenomenon is mainstream when called "suggestion" or "hypnosis." "Voodoo," "faith healing," and "laying-on of hands" are pseudo but the same phenomena called "psychosomatic illness" and "the placebo effect" are mainstream.

Political: "Minorities" in common parlance nowadays doesn't mean any group that is in a minority: it means black, Hispanic, or Native American, but not the multitude of other numerical minorities like Arabs or Jews or Catholics.

Social: During my first visit to the United States in the mid-1950s, I was astonished how similar were the stage goings-on at vaudeville shows (in Nashville) and at burlesque shows (in Detroit); yet the first seemed accepted as respectable, with an audience comprised of well-dressed couples, whereas the second was certainly not respectable, as evidenced by an audience of furtive, poorly dressed unaccompanied males.

However, vaudeville itself was not always or everywhere respectable: the term *legitimate* theater was introduced to distinguish "stage plays . . . from burlesque, vaudeville, etc." (*Random House Dictionary* 1984).

Sexual: "Different sexual orientations" doesn't mean *every* sexual orientation: in most circumstances since the 1990s in the United States, it has meant gay-lesbian-bisexual-transvestite but *not* animal-loving, exhibitionist, pedophile, or other sexual preferences or proclivities.

- *The unorthodox invoke the same overriding principles as the mainstream*—"proof," "openness," "free speech," "justice"—but argue for an unthinkingly literal application instead of recognizing the import of different circumstances. The purpose is to justify the unorthodoxy, not to support the principle.

Scientific: Anomalists are fond of claiming that science should be open to new ideas and facts, should be seeking out novelty rather than resisting it, should not be dogmatic, should publish all data, and so forth, all of them norms or principles often suggested as characterizing good science. But taking those norms quite literally is absurd. Scientists must incessantly make judgments about what it's useful to do and publish, for example. (For detailed arguments and illustrations, see particularly Bauer 1992a:chapter 2; Bauer 1984a:chapter 15.)

Political: Proponents of extreme minority views like to cite relevant constitutional language from the free-speech amendment, the separation-of-church-and-state clause, the constitutional right to bear arms. But it's absurd to suggest that "freedom of speech" covers shouting "Fire!" in a crowded theater or that the right to bear arms entails unlimited freedom to own weapons of mass destruction or makes it improper for schools to refuse admission to gun-carrying students.

Religious: When one who claims direct revelation preaches the use of prohibited drugs as a religious ritual and invokes the freedom-of-religion clause, one can reasonably question which came first, the religious revelation or the desire to do drugs.

Sexual: Some pornographic videos begin with a preamble in which a gravely serious man posed in front of an American flag ponderously invokes the doctrine of freedom of speech. Some pornographic sites on the Internet feature links to groups or Web pages that stand for principled defenses of free expression like "1st Amendment Pledge," "ACLU Freedom Network," "Constitution Society," "Cyber-Rights Home Page," or "Libertarian Party" (for example in 1997, "Red's Freedom Links," <http://pages.prodigy.net/red_daly/redlink2.htm>).

Some who have pushed for nondiscrimination on grounds of "sexual orientation" have consciously employed that principled language as more likely to gain social acceptance than if "gay-lesbian-bisexual" were spelled out, yet the latter is usually the intent, for (as noted above) bestiality, bondage, sadomasochism, pederasty, and other "sexual orientations" are still widely agreed *not* to be among the "sexual orientations" granted nondiscrimination by the wider society.

- The separation between mainstream and unorthodoxy has *connections with or analogies to distinctions of social class.*

Scientific: It's routine for those whose claims are rejected by mainstream science to rail against the scientific Establishment and its entrenched elite. During the nineteenth century, the campaign by homeopathists and other unorthodox healers to be licensed equally with officially accredited physicians was sometimes portrayed as an expansion of the principles of Jacksonian democracy (Kaufman 1971:48).

Political, religious, social: Milton Friedman's economic views were at first ignored or rejected by the disciplinary community whereas a wider public was fascinated. His book *Capitalism and Freedom* "sold more than 400,000 copies even though . . . it was not reviewed in any major national publication" (Brooks 1998).

In matters of politics, almost by definition the mainstream is defined by social class or political power: unorthodoxies are what minorities or lower social classes preach or practice. In religion, the Anglican Church and the nonconforming ones differ more in the class from which their members are drawn than in their religious doctrines; even within the Church of England itself there are similarly differentiated High Church and Low Church (evangelical) persuasions. Though there may be a much larger number of Non-Conformists, the High Anglican Church is the accredited orthodoxy.

Sexual: "Pornography" is reprehensible worthless stuff craved by the hoi polloi and with no redeeming social value; what the elite peruses and purchases is by contrast not pornography but "Erotica and Curiosa—fine and rare material," in the words of one Brooklyn bookseller, C. J. Scheiner, whose ad has long run in the *New York Times Book Review.* The Museum of Sex planned for Manhattan will be "a serious cultural venture featuring exhibitions of erotica and performances by exotic artists. . . . 'The exhibits are by no means pornographic.' . . . [T]he museum was being advised by 'distinguished scientists, sex historians, doctors, psychologists . . . artists, curators and members of the sex industry'" (Allen-Mills 1998).

Advertisements for *The Erotic Review* in the respectable *Times* of London contrast its "discreet" yet "explicit" illustrations with lowlife "ubiquitous commercial exposure," though its own explicit illustrations are "if anything, 'more frank than the images available in newsagents'" (Anon. 1998d). Daphne Merkin remarked (on page 16 of her book *Dreaming of Hitler*), "I learned early on how much more morals-bending was allowed under cover of High Seriousness—Marlon Brando buttering up Maria Schneider's ass in *Last Tango in Paris*—than in the open light of Low Eroticism."

As Camille Paglia noted, it's "elitist class bias" that lauds high art but denigrates pornography: "Pornography is simply art for the masses" (Allen-Mills 1998)—just as demagoguery is politics for the masses, and pseudoscience is populist science.

- *Shibboleths abound* wherever there are strongly held beliefs and clashes of beliefs. I gave earlier some examples in science and in anomalistics. History and social beliefs have their own shibboleths that are accepted unquestioningly just because they fit the politically correct point of view.

  A classic historical shibboleth, or "Tonypandy" (the place where a massacre occurred that in actuality never took place), is that Richard III murdered his young nephews, a tale actually invented by the succeeding dynasty in order to stake its claim to legitimacy (Tey 1951). A contemporary (mid-1990s) instance is that there was an epidemic of racist burning of churches in the American South, when in reality the epidemic was an invention of the media (Leo 1997).

  Those illustrations are at the same time political too, of course. As to minorities or unorthodoxy more or less in general, there are such widely sustained shibboleths as the scientist Martin Arrowsmith in Sinclair Lewis's 1925 novel, eschewing fortune for pure research, or the stereotypical artist starving in the garret—so that any commercial success is taken by some purists to prove lack of artistic authenticity; for example, authentic bluegrass is a pure and utterly uncommercial traditional folk music (Smith 1998).

The latter, like the stories of David and Goliath and of Galileo and the church, exemplify the common anomalistic shibboleth that the unorthodox underdog not only should prevail but indeed will win out in the end.

- *Evidence is often ambiguous or evanescent.* The claim of a conspiracy in the assassination of President John F. Kennedy displays some similarities to the claim that an abduction via UFO took place. Both are much-denigrated minority views—that is to say, a minority view according to supposedly authoritative mainstream experts though fairly widely believed by the general public.

  Much or all of the tangible evidence is no longer available. New evidence is hard to come by, since the claimed events are in the past and not subject to repetition. Many of the witnesses or proponents are people somewhat out of the ordinary, whose credibility is not obviously high. Much of the arguing concerns alternative interpretations of the same facts. Ultimately, one forms an opinion less by reliance on the specific facts of the case than on the basis of one's preferred interpretive framework of the world and of society; and so disputation can continue almost without end.

  The believers form a community of their own with little contact with mainstream academic or public media. If the media do periodically pay some attention, it's not in the interest of clarifying the issues but to use the case as an interesting oddity with which to entertain their audience.

- *Passions run strong even though (or perhaps because) the border between mainstream and unorthodox is not so readily defined.* "Drive the pseudos out of the workshop of science" (Wheeler 1979) is the battle cry of those who want the American Association for the Advancement of Science to excommunicate parapsychology. A Distinguished Professor whose studies were branded "alchemy" was urged by some colleagues to resign while the other Distinguished Professors asked the university to strip his title from him (Begley and Levin 1994; Mangan 1994; Pool

1993b). Over an academic conference that featured sexual topics, calls came for the president of the university to resign (Anon. 1998a).

- *The Internet has been a boon to the proponents of unconventional opinion* across the board—politics, science, sex. It's a nightmare for the upholders of conventionality, and to a lesser degree for the uncommitted who seek authoritative, relatively unbiased information and assessments.

Overlaps of Scientific and Nonscientific Unorthodoxies

Beyond substantive similarities like those just described, there are also overlaps of scientific and nonscientific unorthodoxies in terms of who is interested in them. Apparently there is a substantial audience for minority views in general. Perhaps the cast of mind that is independent, intent on understanding rather than accepting on authority, and thereby resistant to orthodoxy is likely to be so in a range of fields and not just in one; and perhaps those excluded from an activity seek allies and may find them far from rather than close to their initial concerns. At any rate, instances of overlaps are common:

- Anomalistic topics in history overlap inevitably with political or religious ideology, for it is not just a matter of establishing "what" happened but of placing it into meaningful context. This requires interpretation, and that presupposes an evaluative viewpoint, which is at the same time an ideology. Conspiracy theories are instances.
- The Internet is a veritable bazaar of intellectual and other populism. Browse among anomalist topics and often there turn up pages in which intellectual and other unorthodoxies are linked in some way, say, cold fusion and libertarianism (for example, in a 1998 Web page of John Logajan, <http://www.skypoint.com/members/jlogajan/>).
- *21st Century Radio®* pays attention not only to "UFOs and Intergalactic Intelligences," "Ancient Civilizations," and "Paranormal in General" but also to "Politics," "Cultural Heroes," and "Survival Preparedness/Earth and Weather Changes" (Hieronimus 1997).
- Looking into the antifluoridation controversy, I recognize a protagonist (Burgstahler 1988) whom I first encountered in the Velikovsky affair (Bauer 1984a:325).

  Becoming curious about the questionable evidence that HIV causes AIDS, I find myself in the company of individuals some of whom I came to know through activity against political correctness, some with whom I came into contact on issues of grade inflation and educational reform, and others with whom I've crossed swords over creationism or orgonomy.

- Individual books pro or con unorthodox matters offer various mixes. Thus a recent work on "Weird Things . . . Pseudoscience, Superstition, and Other Confusions of Our Time" lumps together Edgar Cayce, UFOs, Ayn Rand's "Objectivism," creationism, Holocaust denial, and more (Shermer 1997). The catalog *Lindsay's Electrical Books* offers not only works on radio circuits and neon signs but also on Tesla's lost inventions; on *Earth Energy—A Dowser's Investigation of Ley Lines* (Lindsay Publications 1990); and on the Lakhovsky multiwave oscillator, "tested by physicians of the time who found that it not only had a 98% cure rate for terminal cancer, arthritis, and other 'hopeless' diseases, but that it could rejuvenate plants and animals as well."
- Seekers of a physical Holy Grail or Noah's Ark can hardly keep their quest separate from religious implications and exegesis. Martin Gardner (1997) suggests that many political conservatives find attractive the notion of "intelligent design" as explanation of the biosphere in opposition to "Darwinism"—but of course there are also many political conservatives who do not.
- Social idealism is not uncommonly an instigator of or collaborator with some things that come to be called pseudoscience.

  Regarding Rudolf Steiner's anthroposophy, it has been remarked that "such occult interests were the rule rather than the exception in the idealistic and reforming circles in which Steiner lived" (Cavendish 1989:236).

  "Mesmerism in America was more than an *avant-garde* intellectual entertainment. It became part of a social movement . . . [of] perfectionist political thinkers. . . . The bond between radical politics and occult behavior is still intact today" (Aveni 1996:174).

  Homeopathy had ties to practical democracy, as already noted, and to social reform against brutal prevailing conventional medical practice, besides its intellectual challenge to "allopathy."
- "Skeptics" or debunkers often have their own preferred alliances too with nonscientific causes. Thus Prometheus Books, founded as an adjunct to CSICOP and related ventures, publishes not only scientistic dogmatism debunking pseudoscience but also secular humanistic antireligious works and books dealing with sexual unorthodoxies, critiques of political correctness, and politically libertarian or neoconservative or traditionally liberal works. The Center for Inquiry deals in "skepti-

cism," "evolution *vs.* creationism," "humanistic ethics," "sexual ethics" (Center for Inquiry 1998).

Commercial Exploitation — My interest in anomalistics brings me solicitations and offers having to do with unorthodox or odd-ball attitudes also in political or religious or social matters far removed from scientific anomalistics; thus the summer 1997 issue of the *Truth Seekers Catalog* has sections on "Conspiracies" and "Spiritual Prophecies and Conspiracies" in addition to a range of scientific anomalies like "ET's & UFO's." *The Exotic Research Report* (*Infinite Energy* 1996) combines alternative energy, alternative medicine, "Legislation Affecting YOU!," weather control, Gulf War syndrome, electromagnetic weapons, "New World Order . . . and much, much more!!!"

Fate magazine deals in anomalistics, but the advertisements in *Fate* feature also health and religion and psychic claims and offer money, love, self-knowledge, as well as "scientific" knowledge: "VAMPIRES exist!: Join us! Send $5.00 for lifetime membership"; the University of Metaphysics International offers home study programs to "be a doctoral professional to counsel and teach others how to holistically improve their life, body, mind & soul"; Finbarr International invites, "Invoke the 9 secret names of God and receive untold blessings and riches"; and much else. There could hardly be a clearer indication or summary of how intellectual anomalistics gets tied in with all sorts of wishful ambitions.

Noncommercial Exploitation — Commercial exploitation is not the only sort of exploitation. The high status of science causes individuals and groups of all sorts to portray themselves as "scientific"; even religious fundamentalists appeal to *scientific* creationism (Bauer and Huyghe 1999). Anomalist claims are used in the service of political, religious, or other ideology. The very subject matter of anomalistics often lends itself easily to such exploitation: for example, ufology could hardly avoid considering the possibility of government cover-ups, and that attracts some people interested primarily in conspiracy theories and only secondarily in UFOs.

Here are a few other examples of science or scientific anomalies misinterpreted or misused for ideological purposes:

- The insistence that "yellow rain" in South-East Asia was biological warfare waged against American troops (Ember 1984). Years of actual scientific investigation of the matter by private dissidents was needed to provide the compelling proof that the alleged toxic material was actually feces from bees (Anon. 1989a; Anon. 1989b; Ember 1986; Marshall 1986).

- Lyndon LaRouche seized on cold fusion as an instance of the technological ineptness of our form of political government or management.
- Afrocentrism, primarily a cultural and political ideology, offers as "scientific" support historical "facts" that are repudiated by the historical profession as a whole (Lefkowitz 1996).
- Some radical feminist groups have sponsored the use of "scientific" techniques repudiated by disinterested members of the relevant professions, as for example "facilitated communication" (Burda 1997) or hypnotic recovery of memories of sexual abuse.

Strange Bedfellows — There are few if any necessary connections among individual anomalies, and yet there is quite frequent overlap among various sorts of scientific and nonscientific unorthodoxies. This is illustrated by the wide variety of allegiances one encounters. The "intelligent design" movement includes both "Young-Earth" and "Old-Earth" believers (assigning the Earth ages of about ten thousand years and about four billion years, respectively), which might seem, to those who hold neither, to be totally incompatible positions.

Battling against pornography, radical feminists find themselves allied with arch-conservatives. Battling for other forms of sexual unorthodoxy—gay-lesbian-bisexual life styles—the same radical feminists are at odds with the same arch-conservatives. But confronted by those logical extremists who point out that "sexual preference" surely includes pederasty and sadomasochism, some of the radical feminists discover themselves to be after all on the same side as the arch-conservatives.

Striving against political correctness, I find myself allied with reactionaries, neoconservatives, and traditional liberals. On many issues relating to trashy pseudoscience, I applaud CSICOP's activities; when it comes to disinterested pursuit of unorthodox knowledge, CSICOP is anathema to me.

The variety of such connections and disconnections has at least one moral: ascribing guilt through association in these matters must often be unwarranted. Beyond that, these varied associations result in a bemusing situation:

A, B, and C agree with me over HIV/AIDS. They're sensible and rational on that matter, indeed unusually so since they are with me against the conventional wisdom. Yet they're wrong about intelligent design and orgone energy. Other people think as I do on a number of scientific issues but accept that HIV causes AIDS and dismiss the notion that Loch Ness monsters exist. How can they be so right on many things but wrong about those?

Yet those who disagree with me are surely just as puzzled that *I* fail to agree with *them.*

Is it then at all possible that *I* may be wrong on one or another of those issues?

Thus anomalistics can teach humility just as science can, or religion.

Grains of Truth

In controversies over nonscientific unorthodoxies perhaps even more than in anomalistics, there is the tendency to polarization and an "either-or" attitude on both sides. Just as in anomalistics, this does less than justice to the issues. Knowledge seeking requires that people be free to pursue ideas that others regard as misguided.

Just so in society at large. As the First Amendment recognizes, "political extremists pose less of a danger to the open society than do government actions that restrict civil liberties in the name of protecting us from extremists (such as the Omnibus Counterterrorism Act) or from ourselves (such as the Communications Decency Act)." The "Communist Party USA was the first political party to call for complete racial equality" (Miele 1997). That the Party may well have been doing that for ulterior purposes, and that a Communist USA might have been just as racist as a capitalist one, does not gainsay the fact that the Party was making a valid substantive point at a time when the mainstream needed instruction on that particular point.

To separate the grains of truth that there might be in anomalist matters from the large amount of chaff offers pitfalls galore. Before President Nixon's trip to China in the early 1970s, Americans wanting to learn about acupuncture had no obvious authorities to guide their exploratory steps. Nowadays one who wishes to go further in studying the apparently solid, often beneficial, and a priori *reasonable* psychological-combined-with-physical therapy pioneered by Wilhelm Reich meets the same quandaries—some of the most prominent "orgonomists" make patently ludicrous assertions about matters of physical and chemical science, yet it is only from them that one can learn what Reich's therapy actually consists of.

Such quandaries do not make a matter unworthy of study.

Belief and Action

Ingesting too much vitamin A can cause damage; but the belief that large vitamin A supplements are beneficial does not of itself cause that damage.

Believing something does not necessarily lead to corresponding actions; nor does *not* believing mean that corresponding action will *not* be taken. Beliefs and actions are not inevitably linked, they are not cause-and-effect; that is why social science differs from natural science. There are quite a number of things that I believe to be worth doing but that I do not do. I do believe that being relaxed is good for my health, yet I relax very little. There are also things that I believe to be wrong that I nevertheless commit. And some of my actions stem from tentative beliefs or even beliefs held with ironic skepticism. I don't really believe that Co-enzyme Q-10 will hold off aging, yet I take it religiously.

We're not very good at understanding how crucial and stark the difference is, between thinking or wishing something and actually doing it. Dr. Maxie Maultsby, a wonderfully eloquent and adept exponent of Rational Behavior Training (Ellis and Harper 1975; Maultsby 1975), would say, "Try to get up out of that chair you're sitting in."

As you obediently rose, he would remonstrate: "No! No! I didn't say 'Get up!' I said '*Try* to get up!'"

So much for all the "trying" to lose weight while continuing to snack between meals. If you ought to do a thing, just believing that you should is no substitute for actually doing it.

The connection between beliefs and actions is even less straightforward than Maultsby's demonstration indicates, though, because we often want not just one thing but all sorts of things that are not mutually compatible. While I do believe that relaxing more would be good for my health, I believe it is more important to do "useful" things. Not everyone who believes that wise aliens from a higher plane have paid us a visit will necessarily commit suicide in order to join them, as however did the members of the Heaven's Gate cult. Not everyone who believes that Allah recompenses those who kill infidels spends time killing infidels.

A given religious belief can spur entirely various actions. Some atheists like to cite examples of the harm that religion has caused: pogroms, the Crusades, the Inquisition. But that Christ is held to be the redeeming Son of God does not inevitably entail Crusades or torture of heretics. One can equally point to the inspiration Christianity provided for marvelous good works, say, by Albert Schweitzer or Mother Teresa, or for the Afrikaner Jan Hofmeyr, whose Christian belief caused him to turn away from the orthodoxy of apartheid decades before his peers abandoned it (Paton 1965). Christian sects manage to differ sharply among themselves on many social and political issues: abortion, divorce, women as priests. Some individual Buddhists, Christians, Hindus, Jews, and Moslems find themselves in better agreement with one another over certain concrete issues than with the bulk of those who supposedly profess the same religious faith. It's particular interpretations of beliefs, not the core beliefs themselves, that more directly stimulate action.

Now one might of course seek to argue on grounds of tendencies and probabilities: conceivably, for example, Catholic believers (under the Inquisition) indulged more frequently than Moslems or pagans in torturing those who disagreed with them. However, evidence to such effects is not adduced in these polemics. Rather one encounters superficially plausible but sweeping generalizations, say, that superstition threatens the rationality of society or that scientific illiteracy threatens our democratic way of life. These sweeping assertions are not empirically grounded, however; they are shibboleths, accepted not because they are faithful to the facts but because they fit a given ideology.

In matters of pure belief it's possible to be entirely abstract, intellectual, and rational. Beliefs can be simple, uncluttered by exceptions. But when it comes to taking action, one is forced to acknowledge the existence of clashing beliefs, of emotion as well as reason, of tangible barriers such as the inertia of social as well as physical systems.

Given that believing something is not the same as acting on that belief, we need to consider separately the corollaries of *entertaining* anomalous beliefs and the consequences of *acting* upon anomalous beliefs.

Taking Action

It makes sense to act according to what the demonstrated probabilities are. By definition, anomalistic subjects are longshots, a priori implausible. No one claims it likely that there should be dinosaurs still living in the Congo jungle; some claim, though, that it happens to be so, no matter how improbable it might seem. It would make little sense to take actions, therefore, that could cause significant harm if those dinosaurs were not to exist after all.

But being unwilling to take actions that could have unwelcome consequences is not the same as being unwilling to investigate when no serious harm is likely therefrom. It is entirely reasonable to discourage people from following psychic hunches while at the same time devising experiments intended to demonstrate the reality of psychic phenomena—to follow Flaubert (Shattuck 1998), "One must live like a bourgeois and think like a demi-god!"

Medicine and alternative medicine perhaps illustrate best the dilemmas of unorthodox belief and whether or when to translate it into action. Homeopathy arose in major part as a reaction against the brutal treatments conventional medicine was then offering: bleeding, cupping, venesection. In our day, some conventional treatments for cancer are equally brutal: chemotherapy, radiation, bone-marrow transplants. It is no wonder that alternative medicine seems attractive to those who have been offered those conventional approaches. Moreover standard medicine itself often does not rely on scientific knowledge:

"only fifteen percent of medical interventions are supported by solid scientific evidence" (McCallum 1996:81n49).

Even when there is good reason to disbelieve the theories or explanations offered by alternative medicine, its practice may nevertheless be helpful. In medicine as in technology in general, solid experience is more to the point than contemporary mainstream theories. After all, hypnosis is a preferred conventional treatment to remove warts (Thomas 1979:76–81) without the benefit of a good theory to explain that efficacy. "Therapeutic touch" (which involves no actual touching) has failed to demonstrate efficacy in clinical trials: "The more rigorous the research design, the more detailed the statistical analysis, the less evidence there is . . . [of] any observed—or observable—phenomenon" (McCallum 1996:67). But that only shows a lack of efficacy in general; some individuals might nevertheless have derived benefit from it, just as only a small percentage of people experience side effects from commonly used medications or vaccines. What's certain is that it can cause no harm to the individual who is treated in this manner—which is, far from incidentally, not true of most of the experimental treatments mainstream medicine dabbles in. If I feel better through self-delusion and the placebo effect, at least I feel better—which, after all, is the point of the exercise.

Debunkers of alternative medicine ascribe its successes to the placebo effect. But why should I care how it happened that I feel better, just so long as I *do* feel better? A placebo seems to be somewhat effective in as many as one-third or even more of nonfatal ailments. Moreover it rarely has unwanted side-effects (Harrington 1997; Shapiro and Shapiro 1997).

The choice between conventional and alternative medicine is often not as clear-cut as the debunkers would have it. I'd be very happy if my doctor were to offer me choices of the following sort:

1. Prostate surgery, with an appreciable chance of such undesirable consequences as impotence, incontinence, long-term discomfort, and periodic follow-up surgery.
2. A placebo, offering only 30 percent chance of appreciable relief, but no complications and no long-term discomfort.

Unfortunately, I am offered only the first by my physicians, and the second only by practitioners of alternative medicine.

Taking Risks

Even when the chances are quite slim that an anomalistic pursuit can be successful, and even when lack of success might seem to forebode tangible detriment, some people do choose to take the risk. Steven Roger Fischer and his wife scrimped and saved for decades to give him the time to decipher scripts that conventional scholarship took to be undecipherable, and they don't regret it (Fischer 1997).

Different people judge variously which risks are worth taking. Some believe it worthwhile quite voluntarily to take very great risks, say, in mountaineering. I personally cannot understand the mountaineer when I read about the succession of deaths in the attempts before World War II to climb the north face of the Eiger (Harrer 1962), or that, as of 1996 (Anon. 1997d; Holland 1998), 4,000 had tried to climb Everest, 660 succeeded, and 142 died in the attempt.

Is then an anomalist belief or a religious faith really as dangerous, really so to be avoided, as the debunking sci-cops insist? Compared to mountaineering, which they don't decry? Or compared to using an automobile, which offers an American about a 1 in 5,000 (lifetime) chance of death?

Anomalistics is a matter of intellectual adventuring. Many people—hermits, for example—have gone adventuring in intellectual or spiritual ways that most of us couldn't understand (Zaleski 1997). For that matter, many scientific biographies portray single-minded, driven people for whom intellectual safety was no consideration at all. And humankind as a whole has benefited greatly from the risks that artists, explorers, scientists, and others were willing to take, risks that the contemporary conventional wisdom generally judged misguided.

Jesper Parnevik is a highly successful golfer. Would he be more successful, or a happier or better person, if he did *not* believe, as he in fact does, in reincarnation? Or if he did *not* use, as he in fact does, volcanic sand as a dietary supplement? Or if he had *not* replaced his silver dental fillings with porcelain "to improve his karma" (Bonk 1997)?

These rhetorical questions suggest that we cannot in good faith insist that an individual's anomalous belief, even when acted upon, leads to harm. Or even if it should, that it leads to more harm or is more likely to lead to harm than socially acceptable beliefs like the desirability of being the first person to climb the north face of the Eiger, or than the practical utility of driving about in cars. Yet loud voices continue to insist that anomalous beliefs are significantly damaging. Sometimes the danger of harm is allegedly to individuals. Sometimes it is supposedly to society at large. What real merit is there in those pleadings?

Does Anomalous Belief Harm Individuals?

Where individuals are concerned, commonly cited illustrations of harm include the failure of Christian Science parents to provide needed conventional medical care for a child. This is harm resulting from actions, it should be noted, not from the beliefs themselves; some avowed Christian Scientists nevertheless do call on regular doctors at times—most humans do sometimes sin against their religious beliefs. To repeat, any given belief that one holds doesn't inevitably lead to the logically consequent action.

To get a scientific answer to the question, "Is anomalous belief harmful to individuals who hold them?" one would need to compare the consequences of all actions allegedly stimulated by anomalous beliefs with the consequences of all actions supposedly stimulated by mainstream beliefs. That remains to be done, and in the meantime anecdotes can be offered on all sides. I found heart-rending the feelings of the Christian Science mother who told (at the First CSICOP Conference in Buffalo, New York, October 1983) of the death of her child from appendicitis. But I also find heart-rending many other instances where parents have lost a child and find reason to blame themselves for it. I cannot conclude that Christian Science as opposed to the human condition is at fault. Had these parents held other beliefs, would they necessarily have avoided such a tragedy? Does that not depend on what those other beliefs would have been? Is it not an important factor that these parents followed to the letter the teachings of their faith? Perhaps it was the following-to-the-letter more than the specific belief that was most blameworthy? After all, much harm is done by people who "just follow orders," for quite a variety of orders in a variety of situations. Perhaps much more generally damaging than any given belief are the fanaticism and dogmatism that hinder some people from deviating in any situation from rigid ideologies.

But, of course, even undeviating adherence to fixed belief does not automatically cause harm. When individuals forsake responsibility for their own actions to follow the lead of some ideological guru, almost everything then depends on the guru and the ideology. People who follow orders don't always come to grief nor do they always bring grief to others. Among religious groups, there are monastic orders whose members live apparently contented lives, and the organizations thrive over many generations. One doesn't have to give credence to Mormon or Hutterite beliefs to recognize that many members of those societies enjoy a high level of psychological, social, and physical well-being. Following a guru may sometimes lead to tragedy, as at Jonestown, but often it does not.

Anomalous beliefs can bring quite tangible personal benefits. My own fascination with Loch Ness monsters, leading by the early 1970s to belief that they are real animals, has brought me nothing but good—quite irrespective of whether my belief is true or not, for that remains to be established.

My fascination with Nessies stimulated me to think and learn about the significance of science in a way that practicing chemistry had never done. It led me to recognize the high degree of human ignorance about significant matters. It led me to the extraordinary insight that I may be wrong even when I'm enormously sure that I'm right. It led me to take an interest in other unorthodox matters, and thereby to learn a great deal about all sorts of enthralling things and to come to know some marvelously interesting people.

Nessie was even good for my career. Having through that become more interested in the nature of science than in continuing research in a subfield of electrochemistry, I wanted to make something like history of science, rather than chemistry, my professional pursuit. That led me to seek an administrative position, as a way station while changing intellectual fields. That administrative job brought quite unforeseen intellectual benefits; it taught me a lot of things I never knew existed to be learned (Martin 1988), like the cultural differences between academic fields (Bauer 1990a,b, 1997; Bauer and Huyghe 1999).

So I've acquired useful knowledge and valuable experience as a result of Nessie-questing, and moreover it's been a lot of fun. Several times a year, over many years, I've talked about Loch Ness monsters to school and other groups, and have enjoyed the lively curiosity that young children bring to everything; and I've benefited from some of the questions they ask, ones that led me to learn about other things that it had never occurred to me to find out about, like "Do fish sleep?"

Many years ago, a Japanese friend observed that I practiced "Nessie-do." He went on to explain that the suffix "-do" (as in "bushido," "judo," "Shinto") was also the Chinese "Tao," signifying way or path (fig. 3). The underlying concept, he explained, is that devoted pursuit of even an apparently limited subject—military arts, wrestling, archery, Nessie—if carried on with sufficient dedication can lead to understanding and enlightenment far beyond the particular quest.

What I've gained through Nessie-do others have derived from other anomalistic quests. The engineer Robert Jahn has given explicit testimony to the deep satisfactions that opened for him as he pursued his quest to characterize apparently psychokinetic influences (BBC Television 1994). I think it likely that many other people get from their particular unorthodox passions what I have from Nessie-do and Jahn has from *psi.*

I personally don't believe that humans have a sixth sense. I don't believe that UFOs are anything but misperceptions. I don't think that Bigfoot exists. But Nessies have taught me that I could be wrong about any one or more of those; and that even if I'm not, there's a great deal to be gained—and nothing that I can see to be lost—by trying to find out for sure.

Figure 3. The Tao of Nessie (Nessie-do)

The "do/to" in "judo," "shinto," and "bushido" is the same symbol as the Tao, which itself signifies moving onself forward intellectually or spiritually. The concept is that *any* dedicated pursuit can lead to wider enlightenment, be it wrestling, religion, military work—or questing for the Loch Ness monster.

Does Anomalous Belief Harm Society?

So an individual's belief on a given matter doesn't inevitably lead to any given action, and consequently not inevitably to harm. Nevertheless, might not a given unorthodox belief prevalent in society at large have harmful consequences?

Just as with individuals, the answer turns out to be, "No, not necessarily."

Yet loud voices warn us that paranormal beliefs do represent a serious threat to society. According to CSICOP, "a lot of money is wasted on horoscopes and clairvoyants, often by people who are least able to afford it. Moreover, some men and women actually base marriage, career and other crucial life decisions on pseudoscientific mumbo jumbo. Or, relying on faith healers or quack remedies, they put off seeing legitimate doctors until it is too late"; "the same as we label packets of cigarettes as dangerous to health, astrology columns should display a proper label concerning their contents" (Otten 1985).

That statement seems to me unworthy of a group that labels itself as exemplifying "Scientific Investigation," on at least two grounds:

1. Is it in fact known how many people do actually base important life decisions on horoscopes, clairvoyants, or the like?

   Can it be shown that people are better off who do *not* base their life decisions on such grounds?

   Does CSICOP wish to assert that we should base all our decisions on scientific rationality? Where then would romantic love figure—should marriages perhaps be arranged by wise elders rather than left to the irrationality of emotional youngsters?
2. Reading astrology columns might be harmful, but if so it is in a very different way than smoking cigarettes is harmful. Treating these pursuits as comparable displays a sorry sense of proportion and a failure to make the sort of distinctions that "Scientific" surely calls for.

   Is any conceivable harm from pseudoscience as great as the potential harm from false beliefs about a range of economic and political matters? Is a ufologist as great a danger to our society as (on one view) a tax-and-spend ultraliberal or (on another view) an extremist libertarian? Is mega-dosing on vitamin C as harmful as following the precepts of Christian Science? Are *any* of those anomalous beliefs as harmful as a literal interpretation of "the right to bear arms"?

Martin Gardner tried to make the case that belief in "fads and fallacies of science" is harmful: "Thousands of neurotics desperately in need of trained psychiatric care are seriously retarding their therapy by dalliance with crank cults" such as dianetics or orgonomy (Gardner 1957:3, 6). But is that "trained psychiatric care" always so effective in helping neurotics? When some of its practitioners favor treatment with drugs, and others favor psychotherapy (by several distinctly different approaches), and others again deplore both of those? When a psychotherapist can write a book (Ellis 1989) about why so many therapeutic approaches (other than his very own, of course) don't work? Some people call Carl Jung the guru of a cult, a prophet—"The Aryan Christ"—rather than a psychiatrist: yet, "'The Aryan Christ' fails to convince me that any serious damage has been done by a few people's adherence to an obviously wacko creed" (Kendrick 1997). Of course one can cite mass suicides at Jonestown and Heaven's Gate as harm done by cults; but what about the suicide of a psychiatrist's patient that another psychiatrist blames on the therapeutic approach used (McHugh 1994)?

Gardner's claim here underscores that the properly valid comparison is not between unconventional beliefs and the ideal rationality that some would ascribe to science, but between the actual avenues open to people in everyday life—say, between advice from accredited psychiatrists and advice from alternative sources. The success rate of conventional psychiatry is sufficiently low that there is no obvious global answer, I would suggest. There are probably a number of practitioners of alternative therapies who have helped a larger proportion of their patients than have quite a number of accredited psychotherapists, as no doubt there are many practitioners of alternative approaches who help fewer of their patients than many accredited psychotherapists do. Nor is the line clear in psychotherapy between what is properly trained mainstream practice on the one side and improperly trained or fallacious on the other; thus, many of the psychologists who have hypnotized their clients into falsely recalling themselves to have been abused by a parent had been properly credentialed and accredited by mainstream institutions.

It may be salutary to recall that during cholera epidemics in the nineteenth century, the death rate of patients treated by homeopathy was sharply lower than that of patients treated by practitioners accredited by the American Medical Association. Not that the homeopathy was so effective, but it was less harmful to its recipients than the treatments then considered proper by official medicine; a leading physician acknowledged later that homeopathy had demonstrated that nature itself could heal better than the "heroic" measures medicine advocated in those days (Kaufman 1971:46, 110).

Gardner pronounces harmful even fallacies that have no implications for health, like "the treatises on flying saucers" or Velikovsky's notions. "It is hard to see how the effects can be anything but harmful. Who can say how many orthodox Christians and Jews read *Worlds in Collision* and drifted back into a cruder Biblicism because they were told that science had reaffirmed the Old Testament miracles? . . . [I]t is easy to forget how far from won is the battle against religious superstition" (Gardner 1957:6).

Surely Gardner overestimates the influence of books like *Worlds in Collision.* Our beliefs tend to withstand all sorts of contradictory evidence and emotional enticements. But beyond that, Gardner presumes dangers associated with "religious superstition"—by which he means religious beliefs held by many intelligent people. Certainly they are held by more people than the number who hold Gardner's views on the matter. What gives him the warrant to pronounce the others' views harmful?

Gardner asserts that belief in UFOs causes us to distrust the American government (Gardner 1957:6); but there are other grounds for distrusting the government than suggested UFO cover-ups. Moreover, according to Gardner (1957:7), "a renaissance of German quasi-science paralleled the rise of Hitler" (Gardner 1957:6, 7). But Gardner's no doubt careful choice of the word "paralleled" concedes that such a correlation is no proof at all that one is cause and the other effect. Economic and historical political factors surely contributed more to the rise of Hitler than any dabbling in astrology or other pseudoscience did; indeed one might plausibly ascribe both that dabbling and Hitler to those broader social factors. That there was something rotten in the state of the Weimar Republic, hardly anyone would dispute; but exactly *what* was rotten *is* disputable—"There are many reasons for the fall of the German Republic. . . . But if you happen to meet abroad the men who occupied the decisive positions at decisive moments, you have your answer, you need seek no further for reasons for the collapse" (Mann 1938, quoted in Craig 1990:16).

I am in large agreement with Gardner over what the good society would look like. I too dislike shoddy thinking, gullibility, and charlatans. But I don't see books like *Worlds in Collision* as significant factors in producing shoddy thinking or gullibility. To my mind the single most important factor is the failure of American schooling to teach substantive material and critical analysis of it, choosing not to nurture intellect but instead to play to emotion—not only with the children themselves but also with parents, the public, the media—to the extent that even in college "education" there is more emphasis on entertainment and social engineering ("diversity") than on intellectual rigor. In my view certain textbooks of "education" are more damaging, intellectually and in their practical consequences, than *Worlds in Collision* ever was or could have been.

I believe it important to resist not only religious superstition but also scientistic superstition, the notion that science and only science has all the answers. Some debunking of "pseudoscience," though, plays into that superstition just as much as works like *Worlds in Collision* play into religious superstition.

The debunkers' case that pseudoscience is a clear and present danger to society is based on only superficially plausible grounds:

1. Anomalous beliefs are false, untrue.
2. Beliefs determine actions.
3. Actions based on false beliefs cause harm.

On reflection, none of those turns out to be correct. Anomalous beliefs are by no means automatically false: a number of once-anomalous beliefs have turned out to be correct and the majority view wrong. As discussed earlier, human actions do not follow from human beliefs as effects follow causes. And I have just argued that actions supposedly based on false beliefs may be less, equally, or more harmful than those supposedly based on conventional beliefs; circumstances alter cases.

If false belief led inexorably to harmful action, we would be in even deeper trouble than we are. We have, after all, no guarantee that any of the beliefs we hold are *not* false. It is fortunate indeed that the consequences of a belief do not correlate neatly with the truth or falsity of that belief.

False beliefs can stimulate actions that produce agreeable results. Precepts of sympathetic magic, or the related alchemical notions of connection between the individual and the cosmos, between animate and inanimate, are false (so most of us believe nowadays). Nevertheless those false notions led Paracelsus to the beneficial introduction of metals and minerals into the practice of medicine, for example, mercury as a cure for syphilis. Until we know literally everything, humility is in order. Jacob Bronowski (1973:139–40) observed that sympathetic magic

> seems a terribly childish theory today, a hodge-podge of fables and false comparisons. But our chemistry will seem childish five hundred years from now. Every theory is based on some analogy, and sooner or later the theory fails because the analogy turns out to be false. A theory in its day helps to solve the problems of the day. And the medical problems had been hamstrung until about 1500 by the belief of the ancients that all cures must come either from plants or from animals—a kind of vitalism which would not entertain the thought that body chemicals are like other chemicals, and which therefore confined medicine largely to herbal cures.

Could a belief in voodoo have any redeeming features? Yes, according to Joseph Needham (1969:162):

> If I believe that by taking a wax statue of the chairman and sticking pins

> in it I can cause him evil, I am adopting a belief for which there is no foundation, but I do at any rate believe in the efficacy of manual operations, and science is therefore possible. The rationalist theologians and the Confucians were against using their hands. There has . . . always been a close connection between this rationalist anti-empirical attitude and the age-old superiority complex of the administrators, the high-class people who sit and read and write, as against the low-class artisans who do things with their hands. *Just because the mystical theologians believed in magic, they helped the beginning of modern science in Europe, while the rationalists hindered it. . . . We cannot say that all through history rationalism has been the chief progressive force in society* [emphasis added].

False beliefs, then, do not inevitably have harmful consequences in action. Nor do right beliefs guarantee lack of harm; as ought to be well known, the road to hell is paved with good intentions. The results of actions and the reasons for taking the actions may not be connected logically or morally. Therefore it makes more sense, it's more useful, to assess the merits of actions in the light of their consequences rather than in the light of what led people to take the actions.

When it comes down to it, much of the polemics between true believers and true disbelievers is just warfare between opposing ideologies. Perhaps it's part of what seems to be instinctive human xenophobia, to want everyone to believe exactly as we do quite irrespective of what actions, if any, might or might not follow therefrom. I have never forgotten the instructive fact that one of the bitterest intellectual arguments I ever engaged in, over whether or not (in the mid-1950s) nuclear tests should be stopped, was with a man who agreed with me that they should be stopped. We had, however, distinctly different reasons for wanting that action to be taken. We argued bitterly because our worldviews clashed and we each thought everyone ought to share our own; that we would act identically on the point supposedly at issue was less important than the difference in underlying beliefs. So too Martin Gardner just can't understand how Bauer, sensible about Velikovsky, could possibly take Nessies seriously; Bauer shouldn't believe that—just because Gardner doesn't (Gardner 1985). CSICOP desperately wants the whole society to believe as it does.

"Skeptics" groups and individuals of that ilk are right, that science is our present best guide—but only on scientific matters. Untold millions of human beings over many millennia have managed quite well without scientific knowledge, the scientific method, or the philosophy of science. In even the most advanced societies, most people manage quite well despite holding opinions that are false: about ethnicity and race, for instance, or about connections between religion and cosmology. Personal convictions can be innocuous, or even personally useful, even when they're philosophically or scientifically insupportable. People come to no obvious harm even though they act on the belief that driving is safer than flying, or that trouble awaits those who see black cats or who walk under ladders or who break mirrors.

Those trained in science do not necessarily behave in socially useful rather than harmful ways, nor in individually useful rather than counterproductive ways. Some who become overly caught up in queer anomalistic matters may also behave queerly in everyday matters; but then so do some in the mainstream. The stereotype of the absent-minded professor captures the truth that some accomplished intellectuals get so caught up in abstraction that they become inefficient or dysfunctional in the real world. What is appropriate speculation in philosophy or excellent science-fiction may not be appropriate as a guide to everyday action. When true believers in UFOs (leaving aside what that might actually mean) seem to get so engrossed in it that they neglect other aspects of life, is it the fault of the anomalistic subject or belief? Any more than it's the fault of chess or bridge, that some people come to regard it as the only thing worth doing in life? Or any more than it's the fault of mathematics that some of its practitioners shouldn't be allowed on the street by themselves?

Anomalous beliefs are not necessarily false; and even when they are, they are not necessarily damaging. Harm is caused by actions, not by beliefs. The distinction is truly a crucial one. Take the First Amendment of the Bill of Rights of the American Constitution: individuals are free to believe and to say whatever they like, because beliefs and words do not cause harm. Exceptions are made only where the probability is very high that the speech will directly lead to harmful actions: one is *not* free to shout "Fire!" in a crowded theater. But the Communist Party could not be banned just for preaching revolution. A number of studies over the years have tried without success to demonstrate that pornography per se is harmful, and so pornography in general cannot be banned. It would be banned only if its specific forms bespeak tangible harm, for example, where children are used in producing it.

True disbelievers damage their cause when they fail to distinguish between trashy pseudoscience and serious anomalistics, between probably false beliefs on differently important aspects of life, and between beliefs and actions. Thus Arthur C. Clarke would not be overly concerned even if, say, Uri Geller were a total fraud. As Clarke wrote of Geller, "He provided the world with much harmless entertainment during a gloomy period of history, and made some distinguished people look very foolish—always a most enjoyable spectacle from the sidelines. For a superb example of such a group, preserved like flies in amber, see *The Geller Papers,* edited by Charles Panati. Many of the authors of this astounding farrago must now wish that the entire edition had dematerialized" (Clarke 1984a:11).

Distorting Knowledge-Seeking

While anomalistics cannot be shown to be harmful, the trashy exploitative stuff—such as healing magnets—can cause damage in the same way as other confidence trickery and commercial cheating do. From an intellectual point of view, while the trash may not in itself addle brains, it certainly does nothing to *un*addle them. I agree with the debunkers that it's a pity when people think crookedly rather than straight, but that's no fault of parapsychology or ufology or cryptozoology.

It has been suggested that rubbishy pseudoscience has an unfortunate influence on science fiction (PN 1979) and thereby on scientific illiteracy, since science fiction has become a significant source of scientific knowledge for some of the lay public; and some SF magazines feature articles purporting to be factual but featuring such imaginary things as psionic machines. Yet the answer, in my view, is not a crusade against populist pseudoscience but better education of children to think for themselves. More intellectually harmful than unconventional beliefs, to my mind, are fanaticism; dogmatism; the failure to make necessary distinctions; the failure to recognize the innumerable nuances and shades of meaning and mixtures of views that make every person's beliefs unique, internally inconsistent to a greater or lesser extent, and frequently poor predictors of behavior. The crusaders against pseudoscience behave more like fanatical evangelists of a scientistic religion than as dedicated public educators. Better education and clear thinking are helped neither by the scicops nor by the other extremists, the New Age mystics and their cousins the postmodernists and deconstructors.

Trashy pseudoscience does intellectual damage perhaps most directly to anomalistics itself, because serious investigative knowledge-seeking about anomalies is held guilty by association. Irresponsible peddling of improbable notions interferes with responsible contemplation of far-out possibilities. Clarke writes:

> Sometimes it doesn't really matter, and there may even be occasions when the most rubbishy of books may open up a mind to the wonders of the universe (as bad science fiction can also do). *But there are times when real harm can be done to serious and important studies, or to the elucidation of genuine mysteries, by the activities of frauds, cranks and hoaxers.* Thus the idea that earth may have had visitors from space is a perfectly reasonable one; indeed, I would go so far as to say that it is surprising if it has *not* done so during the past billions of years. . . . Un-

> fortunately, books full of faked "evidence" and imbecile archaeology have scared serious researchers away from the field. (Clarke 1980:11; emphasis added)

In similar fashion, it may be that Velikovsky's scientifically incompetent "catastrophism" made astronomers less willing to consider the superficially similar ideas ventured by the entirely competent astronomers Clube and Napier (1982).

Acquiring Anomalous Beliefs

A common rhetorical gambit in debunking is, "How could anyone believe that?!"

But human beings believe all sorts of things, and they acquire beliefs for many reasons other than that they are true or well supported by evidence. Whether we're inclined to entertain unorthodox views or not to depends on our upbringing, on our schooling, on a host of early experiences. No matter whether we are debunkers, gullible believers, or middle-of-the-road skeptics, none of us can in unencumbered freedom choose beliefs solely on the basis of evidence and logic.

Schools are often described as being in the business of "educating," but it's a tricky question, whether schools educate or indoctrinate. Education surely ought to mean helping students become better at thinking analytically and critically even if they arrive at different conclusions than do their teachers. But we tend to call it education when the conclusions please us and indoctrination when they don't.

Students tend to accept the views of teachers and textbooks, at least at first. When it comes to anomalistic subjects, several teachers have claimed success in helping students to become more critical thinkers, using as a criterion that belief in anomalies was less at the end of the course than at its beginning (Gray 1984; LeGrand and Boese 1975; Tobacyk 1983). Perhaps that was only a move toward the teacher's views, however, not toward critical thinking; moreover the effect seems to be of short duration (Gray 1984, 1987). After taking a course in archaeology, students were just as likely to believe extreme claims related to archaeology as students who had not taken the course, unless those specific claims had been part of the course content (Feder 1986). Students hadn't learned to think critically in general; they had merely been taught that certain ideas were not to be credited.

Each age and each culture has its Zeitgeist to which all bow to varying degrees. Modern Western society infuses its members with some mixture of the Romanticism and the Enlightenment rationalism that have been features, alternately dominant, of the West's intellectual life for centuries (Brush 1978; Debus 1978). Knowledge fights over anomalies have been a prominent feature of public discourse in the New Age that came after World War II; ufology began shortly after that war ended, *Worlds in Collision* was a bestseller in the early 1950s, and serious interest in Loch Ness monsters was rekindled at about the same time. Surely there was something in the cultural air to produce this variety of instances of anomalist interest. Again, in the early 1980s were founded the International Society for Cryptozoology, the Society for Scientific Exploration, and CSICOP—a largely believing group, a neutral anomalistics society, and a debunker-skeptic group—again manifesting that this time was opportune or appropriate for concern with scientific mysteries and unorthodoxies.

Also beyond the reach of any deliberate intent in the choice of beliefs are such factors as the sibling structure of one's family. Children seem to grow up with religious, political, and other beliefs that are similar to those of their parents or else with something like diametrically opposite ones. Sulloway has shown how consistently first-born children tend to side with their parents' views while later-borns tend to dissent, and how these tendencies continue through life in the form, respectively, of a generally conservative stance or a generally radical one (Sulloway 1996).

Realizing that intelligent others have reached beliefs different from our own can be a salutary aid to reexamining our own beliefs. The implications of UFOs as real phenomena are worth consideration, I've come to think, just because several people I respect take those claims quite seriously. Again, for me perhaps the strongest evidence that psychic phenomena may be real is that such people as Robert Jahn and Robert Morris have devoted much effort to elucidating them.

Perhaps some people acquire unorthodox belief in a sudden revelation, like Saul becoming Paul. Velikovsky, it seems, had a sudden flash of inspiration (Bauer 1984a:164–65) that physical effects of a comet's approach to the Earth, and other such celestial events, could explain much about human history. To others and I think more commonly, heretical belief comes more gradually. It may start with being asked for a professional opinion; being led by that to doing some reading to give the best possible answer; becoming intrigued by not being able to provide a fully satisfactory answer; proceeding to believe that the matter is worth considerable attention. If one spends enough time being curious about a subject, for whatever reason, then regarding it as substantive may follow almost automatically; my high-school minister used to challenge anyone to read the Gospel of St. John three times without becoming a Christian believer.

I know that in 1958 I thought the Loch Ness monster a hoax and a tourist trap. Dinsdale's book aroused my interest in the early 1960s. It became a hobby, and I read whatever I could find; but it took about ten years, and coming to know Dinsdale personally, and seeing some other photos, to bring me to a conviction that Nessies are real. That relatively slow progression from disbelief or ignorance through curiosity to belief seems to describe the acquisition of unorthodox belief by a number of other people too, Allen Hynek over UFOs, for example. Robert Jahn started into psychokinetic studies through agreeing to supervise an undergraduate research project (BBC Television 1994). Linus Pauling, in response to a question, had once ventured that he'd like to live another twenty-five years in order to see where science was going. A biochemist wrote to advise him that he could live that long if he would take supplements of vitamin C. So Pauling looked into the evidence and found it plausible; over the years his belief hardened until many mainstream scientists came to regard him as unreasonably dogmatic about it. Jacques Benveniste did not set out to provide scientific support for homeopathy. A visiting researcher in his lab presented him with unbelievable results, which Benveniste's assistant then proceeded to reproduce, and Benveniste was hooked. Michel Schiff, outraged by the offensive manner in which Benveniste was attacked, tried the same sort of experiment and was also hooked (Schiff 1995:3).

For those who take a debunking point of view, of course, a continuing investment of time in delving into anomalistics does not always lead to anomalistic belief, though it has happened on occasion (for example, Daryl Bem); it does, however, seem to entrench the conviction that the existence of that belief in others is somehow important and positively dangerous. How else to justify spending so much time on it?

Maintaining Anomalous Beliefs

Whether we come to beliefs quickly or slowly, once we have them we tend to keep them (Marks and Kammann 1980:chapter 12), moreover with increasing firmness. I imagine that few students of psychoanalysis begin with the attitude that anyone who disagrees with their interpretation is just manifesting "resistance," but by the time they're full-fledged psychoanalysts, that stance seems to be common.

It's natural to take it as supporting evidence of being right, if nothing happens that forces us to realize we're wrong. Since we tend to notice what appeals to us and to overlook what we would prefer not to be so, we tend to keep whatever beliefs we happen to have already (Gilovich 1991; Marks and Kammann 1980; Neher 1980; Reed 1988). And as already discussed, false belief isn't necessarily harmful; so acting upon our beliefs often seems consistent with getting what we aimed for, whereby again our beliefs get steadily reinforced even when they're false.

Assessing Evidence

Most heretical scientific claims turn out to be false. Some are simply intellectual trash offered solely for the seller's gain. Even if not, then taking a heresy seriously puts one in the camp of the unorthodox, which brings with it some not always attractive consequences. Moreover, even if we believe we are impressed just by the evidence, in point of fact we're likely attracted also for other reasons, as noted above. And then one needs to recognize the vital difference between indulging one's intellectual curiosity, which presents few hazards if any, and taking tangible actions predicated on the chance that the unorthodox view is largely correct, a course that *may* present substantial hazards.

Even apart from all those provisos, assessing the evidence to reach a reasonably objective opinion over an anomalistic issue is no mean feat. The approach one must use depends on the question being studied. To pursue an anomalist quest in a rational fashion is quite a different thing from being logical in thinking through a scientific question. Rationality in anomalistics is different from rationality in science because hard cases make bad laws—or, a corollary infrequently noted, good laws may not be adequate guides in the hard cases. Anomalistics focuses on hard cases, the mysteries that science is not ready or not equipped to grapple with. In science it makes perfect sense to follow the most plausible path, to stick with the tried-and-true in method, facts, and theories; but that is not possible in an anomalistic quest. A different attitude, a different form of judgment, is therefore appropriate to anomalistics.

Low Probability of Being Right

Anomalistic quests are very long shots indeed. Practitioners of cryptozoology are searching for large creatures unknown to science in lakes around the world and in the oceans; for man-apes in several regions in Asia and the Americas; for still-extant giant sloths, Steller's sea-cows, dinosaurs, and more. Not many of those individual quests are likely to be successful, though one or another of them may well be. By contrast, mainstream scientists looking for new species work where large numbers of unknowns are almost certain to exist, say, in rain forests; they care less what particular ones they come up with than that they come up with something even slightly new. Dewey Larson, who has created his own heretical system of physics (Larson 1965), put it like this: "Research is, in some respects, like fishing. If you make your living as a fisherman, you must fish where you know there are fish, even though you also know that those fish are only small ones. No one but the amateur can take the risk of going into completely unknown areas in search of a big prize. . . . [W]e uncommitted investigators are primarily interested in the fishing, and while we like to make a catch, this is merely an extra dividend" (Larson 1979:x).

That few anomalistic projects will succeed, that the ratio of triumphs to successes is low, is illustrated by the fact that anomalistics always cites the same few cases of vindication: meteorites, giant squid, ball lightning, a few others. I'm reminded of Otis Singletary, who for two decades in his public speeches as president of the University of Kentucky rarely failed to mention the quarterback who had a 3.5 grade-point average and went on from football to medical school. That does happen, but it remains the exception and not the rule, no matter how often the successes are cited.

A Venture for Amateurs

Many anomalists are reluctant to concede the corollary that anomalistics is properly an amateur pursuit (Bauer 1982), as Dewey Larson notes. "Amateur" stands here of course in its original sense of doing something for the love of it, not the sometime secondary meaning of less than professionally expert.

Modern science grew out of natural history in large part through the work of dedicated amateurs, who sometimes became the leading experts—Hugh Miller in geology, say, or Boucher de Perthe in human paleoanthropology (Grayson 1983). Modern archaeology owes much to Heinrich Schliemann, an amateur enthusiast and retired businessman. His heretical quest for Troy was based on reading Homer as factual at a time when scholarship regarded Homer as a poetic purveyor of myth, and Schliemann did find and excavate Troy. In twentieth-century anomalistics, Tim Dinsdale is a good example. His writings over the years show a progression from considerable naiveté about science and the trustworthiness of eyewitnesses to a sophisticated understanding of science's necessary conservatism and the proper role of anomalist researchers.

E. C. Krupp, commenting on archaeoastronomy, put it in a way that applies generally to quests that are not dominated by established disciplines: "I don't think it stands up well to formalization. That doesn't mean its inquiries lack rigor. . . . [But the] . . . investigations [are not] within the context of questions and concerns that are meaningful to the members of mainstream disciplines." As to the value of amateur work, "if the world's inventory of prehistoric rock art is more-or-less saved from destruction, we'll have to credit the amateurs" (Krupp 1998).

Everyone Is Biased

One often hears, "Just look at the facts." The trouble is, even deciding *which* facts to look at involves making choices, generally subconsciously. The Zeitgeist infects us all. Current circumstances determine what criteria we use for looking at things. In the days when science was still largely natural history, as noted earlier, there was nothing irrational about classifying objects by their shapes even as that lumped together—as we now recognize—a type of mineral, a type of fossil, and a type of human artifact. So even the facts we choose to look at do not speak for themselves. Around the end of the nineteenth century, speculation about aliens and possible visitors to Earth focused on "men from Mars" (Bullard 1988). Kurd Lasswitz (1897) had published a very influential science-fiction work about Martian visitors to Earth, influenced no doubt by the "canals" seen on Mars by Schiaparelli and later confirmed by Percival Lowell. After the Soviet Sputnik satellite was launched in 1957, we came to have different expectations more appropriate to the "Space Age" (Cohen 1965). We now look further out beyond the solar system for aliens.

A common illustration of how choices are biased draws on the psychology of perception. What we "see" is actually how we interpret signals from the eye to the brain. Standard examples are ambiguous pictures (fig. 4). Rorschach tests reveal the interpretive tendencies that cause us to "see" something in blots of spilled ink. In the same way, what we "see" mentally is actually how we interpret the "factual" evidence. Without malice aforethought, to the man with a hammer, everything looks like a nail; to a philosopher, theory and method are everything; to a sociologist, everything is socially constructed; to a Nessie hunter, every odd thing at Loch Ness has to do with Nessie.

Anomalistics deals with unsettled questions. The evidence can cut both ways and lends itself to opposite interpretations—just as does the evidence in the scientific mainstream where frontier science is concerned, say, over what caused extinction of the dinosaurs (Donovan 1989; Officer and Page 1996; Stanley 1987), or whether human activities are warming the globe. The opposing interpretive frameworks collapse into one only when (or if) the tangible evidence becomes truly compelling.

Wishful Thinking

We want life to mean something, and that predisposes us to find patterns and connections where there may not be any. We take things personally, we tend to presume that what happens around us occurs for reasons that have somehow to do with us, *because* of us or *for* us. We're reluctant to believe that there may be no identifiable reason for something that happens to us. We want to be in control of our lives, not subject to blind chance or blind fate. So personal experience is a powerful incentive to believe in psychic phenomena, experiences like the phone call from a friend just as we were about to call her. Few people understand the laws of probability well enough to avoid being fooled

Duck facing right—or rabbit facing left

Vase—or left and right profiles

Young lady, facing away—or older lady with large chin and nose, facing forward

Figure 4. Ambiguous Figures

by such coincidences, so "amazing" that they "couldn't have happened by chance."

Just knowing that some coincidences are bound to happen is insufficient safeguard against being stunned by particular ones. I was visiting a friend in London just for the day—a Sunday, so all the shops were closed (it was in 1958). He wanted to show me some slides, but the battery in his little slide-viewer was dead. We went for a walk instead. Lying in a gutter we found a live battery of the right type.

I still find that amazing. Someone who believes in angels or in the possibility of clairvoyance might well regard one of those a better explanation than random chance. And coming upon that battery carried much less emotional charge than, say, an apparent communication from a dying relative would have; how much more difficult it is to dismiss as chance coincidence such emotionally laden experiences.

"Science" finds it hard to be persuasive in these instances because when it debunks them as random coincidence, it takes something away from us; it denies the existence of meaning or purpose in the world. The enthusiastic fighters against pseudoscience believe and say that knowing and understanding science brings psychological rewards in the form of a sense of awe about the wonders of the world. That may be true for them; but for others, statistical explanations and material mechanisms may seem more like taking meaning out of life than adding meaning to it. One might also argue the case that seeing the world only through lenses provided by science is a form of escapism that avoids the real questions of human existence, which *are* about meaning and purpose.

Arguments Are Always Messy

Chapter 3 described some of the features that make anomalistic disputes intractable. The proponents and opponents have incompatible underlying axioms. The arguments are often more over those hidden agendas than over concrete questions. Personal issues intrude on or swamp the substantive ones. Polemic tactics include evasiveness, red herrings, and scoring debating points. Citations given cannot always be trusted, data and quotes are taken out of context, and so on. Logical fallacies and ignorance of important matters get in the way.

All those interfere with assessing the evidence. Of course such confusions are encountered in all human arguments, and in science as much as in anomalistics. On a few points, though, one who wishes to form an opinion on an anomalistic issue needs to judge differently than in a scientific controversy.

Distinguishing Beliefs and Their Proponents — We're generally instructed that arguments should be directed at the substantive issues; *who* makes the arguments is supposed to be irrelevant. In other words one should not argue from authority or ad hominem.

Yet it seems perfectly sensible to take into account a person's earlier accomplishments. Routinely in science, reputation counts for something, in fact for a great deal, and rightly so. But that applies only to normal science; when it comes to revolutionary science or to anomalistics, those bets are off. There is quite a long list of highly accomplished people who committed what their peers called pseudoscience and who by hindsight we agree to have been misguided: René Blondlot over N rays, Barkla over the J-phenomenon, and more (Bauer and Huyghe 1999). But there is also a long list of now-lauded great discoverers who were resisted in their own time despite their good reputations, and who by hindsight we believe to have been right. And there is a goodly list of people with little reputation, or who were outside their own areas of expertise, who made great breakthroughs, for example, Semmelweis (Barber 1961). Einstein was a clerk in a patent office, with a mediocre academic transcript, when he created several Nobel Prize–level theories. And even cranks are sometimes right (and then we call them geniuses).

So the expertise, reputation, and general credibility of the proponents of an unorthodoxy are by no means decisive criteria by which to assess the merits of the case. Something true may be expounded so incompetently by such dubious characters that it falls into disrepute (say, hypnotism because of Mesmer); or people of excellent character and impeccable insight might entirely flub making the proper, logical case (like some of the biologists who debate scientific creationists); or trustworthy people arguing in good faith might make a convincing case for something that happens not to be so.

In anomalistics, bets are more a gamble than a careful estimation of odds, whether one backs a person or an issue.

That One Side Is Wrong Doesn't Mean the Other Is Right — Since intelligent people are to be found on both sides of anomalistic issues, there's likely to be enough substance on each side that neither is likely to be entirely and nothing but wrong or entirely and nothing but right. Like the blind men and the elephant in John Godfrey Saxe's poem, "Though each was partly in the right . . . [they] all were in the wrong."

Unorthodoxies as far-reaching as claims of psychic phenomena, that have been around for some considerable time, are unlikely to yield to a simple resolution. Far more likely is it that comprehension will come from several directions: understanding of historical context, of relevant material phenomena, of cultural norms, of folklore. To figure out what dragons "are," we must know something at least of biology and paleontology, something of folklore in several different cultures, and something about the history of all those things. The truth about UFOs, water monsters, or psychic phenomena is almost bound to teach us more about human thought and activity.

Consider for instance the merman (a male mermaid), featured in myth and folklore, a topic for cryptozoology. It's generally agreed that the merman is mythical, not a real animal. Therefore—so goes a common sort of debunkers' reasoning—the medieval belief that sighting a merman presaged stormy weather was just another superstition. But merman sightings—at least some of them—can be plausibly explained as mirage-effects: "Images . . . severely distorted by strong, non-uniform atmospheric refraction." "The thermoclines that generate merman images are best created when a warm air mass slowly moves over significantly cooler surface air. . . . The amount of optical distortion depends directly on the temperature difference between the two air masses, which in turn determines the strength of the front and the severity of the subsequent storms." "The medieval correlation of a merman sighting with the advent of storms is . . . verified" (Lehn and Schroeder 1981:362).

Even such individuals as Immanuel Velikovsky or Wilhelm Reich, who in many respects typify the notion of "crank" or "pseudoscientist," have been right on some points where the mainstream was wrong. Rudolf Steiner, widely known as the developer of the crank field of anthroposophy (after Annie Besant's "theosophy"), was "a notable figure among progressive educators just after the First World War, and in 1962 there were over seventy schools run on Anthroposophical lines throughout the world" (Cavendish 1989:239). Alfred Korzybski is classed as a cultist or pseudoscientist by some, while others respect certain of his insights into semantics.

Dismissing faith healing and the like as "just" the placebo effect finesses the salient point that beneficial physical change has taken place without physical intervention, which is a mystery remaining to be understood. When debunkers pooh-pooh "spoon-bending" or any other form of psychokinesis by pointing out that "ideas don't move muscles," they neglect the mind-body problem: we do use ideas all the time to move our own muscles (Barzun 1964:293).

Many anomalistic claims, this is to say, encompass some genuine unresolved issues even if the major claim is incorrect in the manner in which it's made. In seeking to crack the mystery, it happens quite often that other, unexpected, things are discovered. What substance is actually there may not be the substance that either proponents or debunkers think there is (as with the *ri*).

It would be easy to satirize New Age notions that one can "immerse oneself" in the environment, entering into some sort of communion with it where previously unrecognized or not-understood things become plain; yet that sort of concentrated focus seems to have produced the first plausible explanation for the shape of the prehistoric Silbury Hill (Devereux 1991). A mainstream archaeologist who had worked for years at the Clava Cairns told of insights he had been able to reach only because he had lived enough time there, at various seasons and various times of day, to appreciate nuances of the color of the sunlight and of the stones (Bradley 1997).

As earlier noted, many anomalistic subjects have more to do with human behavior than with inanimate nature. Folklore and personal experience feature often in anomalistic quests. Therefore, even if the overt claim turns out to be spurious, by that very fact we're then in a position to learn something about how and why human beliefs are formed and transmitted and tested—or not tested. A classic illustration is Hufford's (1982) study of the "Old Hag" type of nightmare, an experience whose essence is the same in different cultures—*something* goes on that's not entirely hallucination. "Sometimes a dream would be an expression of an unconscious desire of the soul, and at other times a message planted there by a guardian spirit. . . . [Those are just] two ways of saying the same thing. If there were such a force as the supernatural, then the soul and the guardian both would be supernatural. If there were no such force, then the soul was the psyche, and the guardian spirit was just the lonely mind's imaginary friend" (Perry 1997:71).

With anomalies, the trick is to discern which few among many implausibilities or apparent impossibilities might in the future come to seem *not* impossible, as a result of the revolutionary discoveries that will spring occasionally from the unknown unknown. If quantum mechanics, why not precognition? The world is full of absurdly weird things, science has already demonstrated: "an occult force linking the Earth to the Moon provokes the tides to ebb and flow, and . . . minute creatures too small for the eye to see are capable of causing diseases. . . . [P]ink worms are dining on solid methane at the bottom of the Gulf of Mexico. And . . . if you send one member of a pair of simultaneously created photons to Bellevue and the other to Bernex (two villages near Geneva), each one knows what the other is doing and itself does likewise. To believe in ESP and little green men seems fairly tame by comparison" (Marshall 1997); "if some savage told us of a magical worm that built a little windowless house, slept there a season, then one day emerged and flew away as a jeweled bird, we'd laugh at such a superstition, if we'd never seen a butterfly" (Becker and Selden 1985:29).

Science groupies and established science in the 1950s pooh-poohed suggestions that rocket launches or atom-bomb tests could affect the Earth's weather—those manmade events were supposed not to be powerful enough. Is there a much better reason in common sense, though, to believe that gases leaking out of air conditioners and spray cans have seriously thinned the ozone layer twenty or thirty miles up, which science now asks us to accept?

Correct Points Don't Justify the Whole — Just as the nay-sayers go astray when they globally dismiss everything about and around an anomalous claim, so proponents fall into the error of taking a partial vindication on one or another point as though it were proof of their whole case. Velikovsky was "right"—

at least qualitatively, if one doesn't go into it too critically—that Venus is hot, that Jupiter is a source of radio noises, that there is a magnetic field around Earth; and indubitably right that Linear B is Greek. But still Velikovsky's main scenario is *quite* wrong (Bauer 1984a). Wilhelm Reich surely had a good point to make, that politics and social norms (for example, about sexual behavior) need to go hand in hand ("Sex-Pol"). His blend of physical and psychological therapy continues to benefit people (Edwards 1967, 1974). But orgone energy remains a figment of his imagination. Chiropractic manipulation can be therapeutically effective, but "subluxations of the spine" are manners of speaking, not the physical realities that chiropractic once claimed.

The proponents of anomalies can be right on all sorts of details that don't really count, and at the same time wrong-headed on the main anomalistic issue. Critics and skeptics may be wrong on all sorts of details but right on the main point. Or, of course, they may not be.

Deciding for Oneself versus Convincing Others — One shouldn't be too ambitious about convincing others. "What evidence would be necessary to convince a complete skeptic—and, conversely, what evidence would disillusion a confirmed believer?" asks Arthur Clarke (1984a:8). The rhetorical question serves to remind us how much the controversies about anomalistics really hinge on hidden agendas.

All experience of arguments over anomalies, as also over matters of religion, politics, and much else, indicates that believers are unlikely to be dissuaded by evidence alone: "It is a matter of Faith, not Reason" (Clarke 1984a:9). Leroy Ellenberger is a rare individual known to me who found himself weaned from Immanuel Velikovsky's claims through studying them and looking closely at the evidence; but for nigh on two decades Ellenberger has made no progress in persuading any of his former Velikovskian colleagues.

It is difficult enough to reach a personal, informed view on matters over which controversy rages; there is little chance that the true believers or true disbelievers can be converted. "The most we can hope to achieve is to make the credulous more skeptical, and the skeptical more open-minded" (Clarke 1984b:237).

Hoaxes Will Be Performed

No matter how many hoaxes may be performed, they do not prove that the whole matter is nothing but hoax. The plain fact is that hoaxes are played by all sorts of people, though the rest of us may not be able to fathom their motives.

Anomalistics attracts hoaxers, no doubt because of the relative ease of getting away with it. Since much of the evidence is itself open to question, and since even the best evidence in anomalistics is open to alternative explanations, identifying something as a hoax calls for actual detection of the hoaxer's activities or supremely lucky judgment. Hoaxes confuse the issue, they do not point to its solution.

Eyewitness and Personal Testimony

Arthur C. Clarke has written about personal observation:

> Anyone who has studied those cases of UFOs *whose explanations have been discovered beyond any possible doubt* will be astonished (and perhaps chagrined) at the mind's almost unlimited powers of self-deception. The inability of eyewitnesses to describe even commonplace phenomena, especially when some emotional stress is involved, has long been recognized—and indeed exploited, particularly in the courtroom. When honest observers can disagree completely on the details of a traffic accident, should we be surprised that a re-entering spacecraft a hundred miles away could be interpreted as a Flying Saucer landing in the next field—as has indeed happened? Or that a random pattern of objects glimpsed under poor lighting conditions (at night, on the borders of sleep . . .) could be mistaken for a dead friend? (Clarke 1984b:239)

Very few people—other perhaps than trained artists or photographers—are able to give an accurate description of something they see but don't recognize. That presents a further, historical difficulty in anomalistics. Descriptions of purportedly anomalous objects cannot be taken at face value—all one can legitimately say is that *something* was seen that was not recognized. Nor can one accept at face value descriptions of happenings that involve not only seeing but active participation.

The many psychological and perceptual pitfalls that make eyewitness testimony fallible are well known to specialists (Gilovich 1991; Marks and Kammann 1980; Neher 1980; Reed 1988). But when we personally experience something remarkable, we tend to forget that seeing for oneself is actually one of the bad reasons for believing something (Bauer 1986:chapter 9).

Hidden Events

Though personal testimony is fallible, reports of unbelievable things cannot be automatically dismissed. There are what Westrum has dubbed "hidden events" (Westrum 1982).

Child-battering is the type specimen. For decades physicians saw evidence of it without being willing to believe it was actually occurring. Once some indubitable cases became widely known, however, the number of reported instances increased rapidly. This is the "report-release" phenomenon, which seems to occur also with such things as sightings of UFOs. Many witnesses will testify only when they do not expect disbelieving dismissal and ridicule; and whether reports are disseminated depends less on the actual incidence of events than on the demand for information, a judgment by the media that publication is called for.

Burden of Proof

The burden of proof is on the claimant.
Extraordinary claims demand extraordinary proofs.

Those aphorisms in anomalistic disputes are not easily applied in practice. Disbelievers say that the proponents bear the burden of proving that their subject is real; but believers can respond that nay-sayers must assume the burden of proving that the "conventional explanations" they give actually hold water. Consider this experience of mine:

I had the extraordinary good fortune to meet the man who was with Kenneth Wilson when, in April 1934, Wilson took the famous head-and-neck photo (Bauer 1986:15, fig.2) that has become Nessie's logo. I'll call Wilson's companion "Smithson," for he never wanted his name made public. Smithson related that he and Wilson had stopped for a rest by the side of the Loch and had been utterly astonished as some splashing noise drew their attention to the water. Wilson grabbed his camera and snapped four pictures. The plates were developed at a pharmacy in Inverness; only two of the four showed anything. Wilson made the two good shots available to the *Daily Mail* with some hesitancy because neither man wanted publicity. Smithson told me how they had seen the neck angle forward (as the second successful but rarely reprinted photo shows, see Bauer 1986:15, fig.2) and then submerge.

At the time I met him, Smithson was already eighty-nine years old and did not care to be besieged by the media; he extracted from me a firm promise not to broadcast what he had told me. But shortly after that, Smithson died; and a little later, on the sixtieth anniversary of the modern-day "discovery" of Nessie, I thought it appropriate to put Smithson's eyewitness testimony into the public record.

When I did, I was met with disbelief. What had I done to check the details of what Smithson told me? Did I tape-record his words, or at least keep notes? Where were those notes? Why should anyone believe the recollections of an eighty-nine-year-old, sixty years after the event? What hidden agenda, rather than any urge to tell the truth, might have caused Smithson to confide this to me?

Here is someone else's personal experience:

Alastair Boyd and David Martin heard from eighty-nine-year-old Christian Spurling that he had helped to fake Wilson's photo; a head-and-neck modeled with plastic wood had been mounted on a toy submarine. Boyd and

Martin waited for the sixtieth anniversary of Nessie's discovery to reveal Spurling's confession, by which time Spurling had died (Boyd and Martin 1994; Martin and Boyd 1999).

But why had the photos first been taken, as Spurling said, with a 35 mm camera and then rephotographed onto Wilson's plates? What about the second photo? Where were notes from the interview with Spurling, or a taped record?

Spurling's story is full of holes, and those questions remain unanswered (Shuker 1995:87); yet no matter the holes, Spurling's "confession" was immediately believed by almost all the media and debunkers. The pundits did not apply to Spurling's tale the skepticism they would have heaped on my rendering of Smithson's tale. The nay-sayers all too often apply the burden of proof only to the anomalist side. Their own explanations frequently leave much to be desired (see, for example, in the Velikovsky affair, Bauer 1984a:chapter 13). They will offer several mutually incompatible explanations for the same anomaly and dogmatically pronounce "it" better than any anomalist alternative.

A powerful reason for finding fault with debunkers who call themselves skeptics is that they are skeptical only in the one direction.

Some Questions Are Not Ripe to Be Answered

"At a generous assessment, approximately half of this book is nonsense," Arthur C. Clarke (1984a:4) acknowledges in the foreword to *World of Strange Powers;* "Unfortunately, I don't know *which* half; and neither, despite all claims to the contrary, does anyone else."

Anomalistics grapples with problems that are not really ripe for solution, that may well remain mysteries for decades or centuries more, where one must be willing to suspend belief *and disbelief* while still somehow respecting logic and the force of evidence.

That isn't easy, for one thing because it goes counter to the human wish to have uncertainties settled. But it is simply the case that some questions cannot be answered without an unforeseeable breakthrough. After the living coelacanth was discovered in 1938, it took fifteen years for a second specimen to be found. Another four decades passed before indications came that these "living fossils" might not be restricted to just one small area off the Comores Islands. Even hindsight fails to suggest anything that might reasonably have been done to bring this knowledge earlier.

When a matter is not yet clear, we have to be content with judging likelihood instead of insisting on a conclusive answer. Clarke (1984b:243) sets his usual excellent example. Acknowledging that his estimates are subject to change at any time, he ranks on a scale of relative plausibility:

• fire-walking ability:	"certainly true"
• apparitions, maledictions, stigmata:	"highly probable"
• dowsing, poltergeists, telepathy:	"possible—worth investigating"
• precognition:	"barely possible—not worth investigating"
• psychokinesis:	"don't know"
• reincarnation, survival after death:	"almost certainly untrue"

Another thoughtful writer on science and anomalies, James Trefil, has a different ranking (Trefil 1978), for example, judging extrasensory perception as more probable than dowsing. Each of us is likely to place a somewhat different probability on each of those things and to rank them in somewhat different order. No matter; the main thing is to recognize that these questions are not ready for a final yes-or-no answer. Therefore we should be willing to remain appropriately tentative and humble about them.

Making up one's mind almost always involves oversimplifying to some degree. Making a decision usually means making a certainty of what is merely a high probability. To avoid premature oversimplification in anomalistics, to recognize that questions remain open, one might recall John Wheeler's pithy criticism of overinterpretation in parapsychology (Wheeler 1979): "Where there's smoke, there's smoke!"

In everyday life, it's a reasonable interpretation that where there's smoke, there's also fire. But in anomalistics, mainstream syllogisms don't apply. Novel *facts* uncovered by unconventional *methods* may require unprecedented *interpretations.* Tim Dinsdale's film shows a huge hump moving rapidly through the waters of Loch Ness, then submerging almost entirely but continuing to throw up a massive wake. That's the "smoke." What type of "fire" can explain it, *what* that hump is, remains to be discovered: none of the suggested explanations is plausible, neither the debunkers' nor the believers'. Each of them leaves too much unexplained and has too many unconvincing features.

The Princeton Engineering Anomalies Research group has a vast amount of data showing perhaps a one-in-a-thousand deviation from normal performance of various devices. That's the "smoke." The "fire" may or may not turn out to be human intention or emotion or will, as the group currently appears inclined to believe. Similarly in parapsychology as a whole, nonchance results are certainly obtained but the conclusion that this means that *psi* exists, that information is acquired by other than the usual sensory channels, would be premature. That question is not ripe for answering.

So Wheeler's aphorism is an apt warning against premature conclusions. It should not, however, be taken to suggest that purely empirical investigation is possible. Studying smoke means paying attention to some things more than to others, and it means seeking the cause of the smoke. One needs to speculate beyond the data in order to decide where research should be directed.

Therefore much of the speculation in anomalistics has to be more conjectural than in science. That's inherent in the nature of the subjects. In the Western past, we used to explain facts of life as the doings of spirits and gods. Continual feedback between observation and speculation, a steady boot-strapping, brought into being modern science. Parapsychology and ufology are still in the early stages of boot-strapping. Wheeler's aphorism should be coupled with a reminder that "unexplained" does not mean "unexplain*able*"; and apparently unexplainable today can turn into explained tomorrow, or a year or a century or a millennium from now.

Lack of Guidance

The emphasis in science is on reliability, and a lot has been learned about how to foster that: communal critiquing and determined conservatism. Anomalistics by contrast depends on creative genius to leap far into the unknown, to devise approaches not previously dreamed of, to think incisively along lines that are precariously speculative, to be both creative and right; and there's a lot that is *not* known about how to do that. So there's little if any general guidance to be had about anomalist pursuits.

No Help from the Experts

Who Is Expert? — There are no experts in anomalistics as such. In any established discipline, most of the professional experts are wary of anomalies. That wariness serves well in normal scholarship and research, but it's inappropriate in anomalistics; it's tantamount to pronouncing the anomalies spurious at the outset. (At the same time, "no proposition is so foolish or meretricious that at least two Nobel laureates cannot be found to endorse it" [Gratzer 1987].)

Anomalistics calls for interdisciplinary and multidisciplinary experience, in the social as well as the natural sciences, even as one cannot foresee which specific disciplines need to be part of the mix. In many cases physical science is likely to be the least relevant of all, for it tends to deal in straightforward cause-and-effect relations whereas such a field as parapsychology, concerned as it is with human behavior, has to deal with probabilistic causation; yet "scientists"—often physicists—are commonly asked by the media to adjudicate.

Medicine, centrally concerned with human behavior, experiences analogous dilemmas. Sometimes the "experts" are epidemiologists, whose expertise is in statistical design and analysis; and they may come up with associations that to the experts in physiology or clinical medicine or toxicology seem unfounded, as for example over Gulf War syndrome. Those most convinced that Gulf War syndrome is real are those who personally experience it, and it carries no conviction for them that "science" cannot find any conclusive physical basis for it. The proponents seek to "give veterans treatment and compensation for illnesses linked to service in the Gulf War 'even if we cannot identify the direct cause'" (Anon. 1997c; Brown 1997), while debunkers find that unwarranted. Over an unconventional cancer treatment, who is expert? The pioneer of an unconventional approach, "Dr. Burzynski[,] . . . the principal investigator in 72 FDA-approved clinical trials," or a debunking guardian of conservative mainstream medicine, "the National Council Against Health Fraud[,] . . . [whose] president, William T. Jarvis . . . has a Ph.D. in Health Education . . . [and] is not an M.D." (Whitaker 1997)?

The domain of anomalies is populated with hosts of people who have no scientific training and who have no scientific scruples; fraud and hoax are a standard feature of the landscape, whereas in science they are rare aberrations. One of the first possible explanations to be considered for an anomaly is a deliberate misleading for fun or profit. Scientists may be quite well equipped to detect illogicality and unsoundness of evidence, for those are always at issue in scientific work; but scientists are not well prepared to deal with clever, deliberate illusionists for whom cheating is a standard and not at all dishonorable technique. In the traditional arts of conjuring, stage magic, or stage mentalism, any means of producing the illusion is legitimate: having accomplices in the audience, using elaborate apparatus, surreptitiously switching things, saying things that can distract or mislead—which in other contexts, of course, would be frowned upon as lying. So what can one make of photos of Loch Ness monsters offered by a stage performer? Tony "Doc" Shiels, the performer in question, has published a book in which he says quite a lot about those photos, yet having read it I felt no better able to judge than before, whether or not he really claims the photos to be genuine (Shiels 1990; Bauer 1993–96). Perhaps for Shiels it wouldn't be "lying" if he gave the illusion of having actually taken the photos—after all, he makes no bones about being a performer, so everyone should be forewarned that anything he does publicly might be "an act."

A recurring illustration of unwitting incompetence by supposed experts is the study of psychic claims by physicists and other natural scientists. They habitually misunderstand their task. Psychic phenomena are evidenced by people, so the relevant experts are more likely to be psychologists and sociologists, conjurers and stage magicians, than chemists or physicists. Scientists tend to assume that parapsychology calls for careful examination of such physical events as the bending of keys, whereas experienced investigators in the field, as well as those knowledgeable in magic and trickery, know that parapsychology calls also—and indeed first—for very careful examination and control of those who cause keys to bend.

There are "experts" on particular anomalistics issues, but they are not reliable guides to the reality or otherwise of the anomaly. The people who know most about anomalies are likely to be either outright believers or dedicated nay-sayers. Rarely will someone largely uncommitted expend the effort to master the material. So the most knowledgeable people tend to be prejudiced toward the anomaly or against the anomaly and just for that reason ought not to be too readily believed.

The Scientific Experts Should Not Be Believed — Scientists should not be believed about anomalistics for the reasons just given, and also because their whole experience in science is to rely on the already solid foundation of reliable knowledge, method, and theory. Any radically new claim, if it isn't clear whether or how that claim can be integrated into the larger corpus of science, therefore encounters from scientists great skepticism and caution. "The same features that make a hypothesis exciting—novelty and unexpectedness—will cause peers to resist the idea because it contradicts prevailing beliefs. . . . Discovery means recognizing something when you don't know what it looks like. . . . [It] depends on an investigator's experience, intuition, and creative insight" (Grinnell 1996). James Trefil (1983) has underscored the extraordinary degree of proof demanded in science to support such claims as to have observed gravitational waves or magnetic monopoles. Even when a claim is less ambitious it is not necessarily accepted at once.

So when it comes to the great anomalies, science is bound to remain aloof. It can rightly say about these anomalies, "That isn't science." When pressed in public arguments, scientists will be tempted to go further and say, "That's *pseudo*science." Yet the occasional anomaly does turn out to be valid; so in any given case, this "expert" opinion should not be immediately accepted.

There's a richly funny literature recounting the host of "expert" predictions that turned out to be absolutely wrong, including assertions about what would never be possible (Cerf and Navasky 1984). Lord Rutherford, the first scientist to accomplish the transformation of one element into another, also asserted that atomic energy could never be harnessed in macroscopic amounts, as did Albert Einstein, whose equation for the equivalence of mass and energy is the very basis for nuclear power (Cerf and Navasky 1984:215). Auguste Comte gave as an example of the sort of thing we would never be able to study, the chemistry of the stars (Knight 1986:83), which just a few years later Kirchhoff was doing by spectroscopy, analyzing the light the stars emit (Sutton 1976).

It is as easy in science as in other human endeavors to be wise after the event. It is also as rare in the sciences as in other human endeavors to be wise beforehand. Excellent arguments asserting why something is impossible turn out to be based on ignorance. After the event, such invalid arguments as the thermodynamic one against polywater seem so convincing only because that part of the conventional wisdom didn't change in that case, over that issue. The history of science reveals, though, that mainstream conventional wisdom has often changed. When it does, the excellent arguments of the rearguard are later recognized as having been fallacious. Only when the mainstream doesn't change do the arguments as to "impossibility" continue to seem well founded. Yet such reprieves may turn out to have been only temporary.

Yet the Experts Are Usually Right — There are scores or hundreds of anomalistic topics. Those that are not downright Zeroth-Kind fallacies are still highly implausible. Granting some approximate validity to our estimates of plausibility, that means only a few will ultimately become established as real.

So the nay-saying experts will in most cases turn out to have been right. But the rub is that we have no really good way to judge on which anomalist topics they will be *wrong.* Mid-nineteenth-century, why would one have put high odds on the kraken, whose arms could envelop ships and drag them into the ocean depths, turning out to be approximately real in the form of hundred-feet-long squid? What possible reason could there have been to take seriously the people who talked of seeing luminous balls rolling down airplane aisles, before "ball lightning" became an established phenomenon (Barry 1980)? Anomalistics involves bets on what is in the unknown unknown, the place from which scientific revolutions ambush us, and no one is qualified to peer into it ahead of time.

The experts are right only on average. That does not make them right in any specific instance. In some few cases the experts are almost certainly wrong. But we cannot know ahead of time over *which* ones they're wrong. Anomalistics, properly pursued, can teach genuine humility.

If the debunkers and the believers were to make wagers, then the debunkers would have to offer high odds to make it rational for a believer to bet on his favorite anomaly. The debunkers would collect very frequently, the believers very rarely. The debunkers would be making a steady though meager living from their bets, whereas the occasional believer would hit a major jackpot.

Authoritative Literature — The literature on an anomalistic topic is of no more immediate help than are the experts. Books and periodicals are often difficult to locate (see, for example, Bauer 1986:49–55). They tend to be highly polarized. Most of the writing is uncritical and by advocates. The debunkers often proffer disdain but inadequate substantive argument.

Observing the Participants — As already argued, the character and general credibility of people engaged in an anomalistic pursuit are not necessarily good guides to whether the subject itself has genuine intellectual substance—whereas in science, the reputations of those involved *are* a good guide. Observing the behavior and tactics of participants in arguments can give good grounds for trusting some and distrusting others. But these will not be conclusive grounds, for as also noted earlier, proponents often do not make the best cases for themselves, and sometimes downright cranks can turn out to have been right. Once again one must not ask for certainty when only slim or high probabilities are on offer.

Debunkers often comport themselves in obnoxious ways. They typically sidestep rather than address the strongest points made by the believers (for this in the Velikovsky controversy, see Bauer 1984a:chapters 13 and 14). Yet debunkers will be right most of the time, simply because only a few of the many contemporary anomalistic claims are likely to have real substance to them. Nevertheless, debunkers are often right for wrong reasons.

The true believers, of course, also sidestep the strongest arguments put forth by the debunkers. But they have no choice, for to accept those strongest arguments would mean to accept the validity of current scientific theories under which by definition anomalies are spurious.

Beware those whose self-proclaimed mission is "to educate the public," to induce scientific literacy in order to banish irrationality, and the like. All too often, they define "education" to mean coming to believe what they themselves believe, and "rational" as thinking as they do.

Self-Help

When it comes to anomalistics, there are no authorities to whom it's reasonable to turn for expert opinion. One can't trust the experts or the encyclopedias, let alone the Internet. Each of us needs to form our own views, by learning as much as possible from the opposing sides and the pundits. Here are some points worth remembering about that.

How Could That Be Known? — The leaders of religious cults are "self-appointed gurus who were never pupils [but] become avatars of the age. They decree what tests need to be passed to attain higher levels. There is . . . 'a certain smell' to it, characterized by the institutionalized absence of questioning along with the idea that knowledge is an easy commodity to come by" (Aveni 1996:311).

Whether one wishes to accept someone else's beliefs is a matter of personal choice. Whether followers of gurus are helped or harmed depends on the guru and on the follower. But if it is knowledge that one is after, then one must ask those who offer assertions, how they came by them. As illustrated in the case of dietary supplements, merely to ask how such knowledge could have been obtained is often tantamount to recognizing that it is opinion, speculation, belief that one is being offered, not knowledge.

Extrapolating beyond the Known — The most reliable knowledge is of repeatable phenomena; every time we add silver ions to chloride ions, a white solid forms. This is operational knowledge. But operational knowledge is a hodgepodge of detail that has no self-evident interconnection. To provide such connections we devise laws and theories. Those are human inventions and periodically need revision. Theories and beliefs are stories, and they are never as reliable as the operational knowledge they were based on.

The judgment that a given unorthodoxy contradicts scientific knowledge is in most cases really a judgment that it contradicts some extrapolation from existing knowledge, an extrapolation using currently accepted theories. But extrapolating (or sometimes even interpolating) is always risky. As one himself famous for a major discovery pointed out, "It is difficult to distinguish the proposals of prophets from those of charlatans" (Van Allen 1986:1075).

An Exception to Every Rule — Philosophers and others have identified a number of useful guides to proper reasoning and the weighing of evidence. In anomalistics, one sees that every rule, no matter how generally useful, may have an exception.

One ought to beware of extrapolating beyond operational knowledge, I've just suggested. But sometimes important discoveries will be missed if one refuses to extrapolate. Watson and Crick—and many other innovators—had to prefer their hunches over the facts as then known, in order to advance insight into nature (Bauer 1992a:21–23).

The most convincing test, scientific or otherwise, is concrete observation. Does a person claim to be able to levitate? Show me! Can pyramids sharpen razor blades? Show me! But this criterion cannot exclude those anomalistic claims that allege avowedly episodic, occasional, capricious phenomena. Take this sort of assertion: *If* a dying person wants strongly enough to communicate with a distant relative, it does sometimes happen via clairvoyance or telepathy or apparition. That is not disproved no matter how often the phenomenon does not occur: "strongly enough" is a limitless excuse. But that it is a limitless excuse, and that the assertion can be neither proved nor disproved, does not make the assertion necessarily untrue.

Scientific mysteries tend to have a relatively short life. For Arthur C. Clarke (1984a), a strong argument against the reality of psychic phenomena is that they have remained mysteries for many centuries. Yet this remains an argument, an inference, and one can propose reasonable counterarguments. The mysteries that science has successfully cleared up are largely those where the phenomena are reproducible. What if *psi* is effective only episodically, as suggested above, or only marginally, say, at a level of once in a thousand times (Jahn 1982; Jahn and Dunne 1987)? Moreover there has been little progress in psychology and sociology either, by comparison with natural science—we continue to appreciate the wisdom and insights into human behavior of playwrights, poets, politicians, and warrior-leaders of a couple of millennia ago. The relation between electromagnetism and life has remained mysterious in important respects over centuries of study. So our general, on-average experience that mysteries get cleared up in a matter of decades is no assurance that *all* mysteries, in particular those having to do with human behavior, can be resolved in our own time, or for that matter in centuries or millennia.

What Is the Division of Opinion? — Though there are no true experts in anomalistics, and nothing can safely be taken on authority, still one can get useful leads by noting who, and how many, believe what.

When a popular magazine reports the results of a Gallup poll, the information is rather like the advertisements that tell us of "nine out of ten experts": the information may be crucially biased by how the sample was drawn and how the question was asked. One would like to know what the division of opinion is among people whose opinions should carry rather different weights, and how opinion has changed over time. Unfortunately such information is rarely if ever available. When it is, it can be useful:

A study of belief as to Loch Ness monsters (Bauer 1987) showed that during the 1950s there was a pronounced shift away from scoffing and toward taking the possibility seriously; that there was a high degree of open-mindedness in the pieces appearing in the mainstream scientific literature (though such pieces were relatively infrequent); that the more expert the opinion (taking marine biologists as the most expert), the more credence was paid to the possibility of Nessies' being real. None of those in itself entails a decision, of course; less scoffing in the 1950s could be due to evidence obtained, or to an increase in public gullibility, or to other things; that marine biologists are more open-minded than others might reflect the generally greater state of uncertainty in their field rather than a greater plausibility of Nessies; and so on. Still, the more one knows about everything related to a topic like this, the more satisfyingly informed an opinion one is able to reach.

Does the Topic Have Strong Intellectual Connections to Any Others? — When someone calls a thing "scientific," apply the criterion of connectedness. Assertions that are not closely bound to current scientific knowledge and to work being done currently in the scientific community are not thereby always to be dismissed; but they certainly should not be called "scientific." Those who do so call it are trying to gain credibility in an unjustified way.

Does the Topic Have Political or Ideological Connections or Implications? — Again, if it does, that proves nothing as to its intellectual substance, but at least one may be forewarned about strange bedfellows with whom one might seem to be making common cause, and with whom one will certainly be accused of making common cause.

Humanity's Ignorance and Personal Ignorance — Ignorance plays a huge part in anomalistics. Realizing that may not provide any specific guidance, but it is a caution that, no matter what opinion one reaches or how, something might pop out of the unknown unknown to prove it wrong.

The knowledge needed to form a truly well informed opinion will not all be available, since the question remains open; either the knowledge simply does not exist, or it does but we are unaware that it does and that it would be relevant. How could a ufologist know to consult a spider expert about "angel hair"? How could seventeenth-century paleontologists have realized that belemnites, sharks' teeth, and axe-heads are not similarly conical objects but essentially different ones? How many people know that it takes only a little persistence and practice to balance almost any egg on its large end? How many people are aware that car windshields soon develop a host of microscopic pits that one does not normally notice unless caused to do so by an unidentified flying object overhead (Medalia and Larsen 1958)?

There is a very high level of ignorance about how science works, how authoritative it can be, what makes something science. Not much about the history of science is widely known, nor about philosophy of science, which does *not* characterize science as "using the scientific method" or as dealing only in falsifiable theories, as the two most popular misconceptions continue to assert.

Perhaps most serious, in trying to come to grips with an anomalous claim, is our general ignorance of probability and statistics. We tend to overestimate greatly the odds against coincidences; to take things personally instead of recognizing the workings of chance; to misunderstand what "randomness" means. Most of us cannot calculate correctly the odds for many everyday occurrences. Few know about regression to the mean and its significance. We accept a "statistically significant" result as true, when actually it is wrong at least 5 percent of the time (if the typical 95 percent probability limit was used). Most of us do not understand that huge odds against something's happening by chance might not be so impressively huge if viewed from a Bayesian perspective. We are unaware of population stereotypes: that given, say, a choice between the colors blue, green, and red, we will *not* find about one-third of those asked choosing each one.

Degrees of Improbability — Anomalistics deals in the inherently improbable. History tells us that some small proportion of what was once considered improbable later became accepted common sense. It is unlikely enough that we can select a present-day anomaly that will turn out to be substantively true; it is astronomically more unlikely that we can select two, let alone more. One might then conclude that it is rather safe to reject the claims of those who are omnivorous anomalists.

Alas, even this conclusion needs slight modification. A debunker might point out that surely not both precognition *and* telepathy will turn out to be real abilities; but one might reply that these are not two separate things but merely different aspects of the single facility, *psi. Many* things that seemed impossible a thousand years ago are common practice now.

What seems probable to one person is not so to another. Bayesians can differ with non-Bayesians as to whether or not "chance" is a likely explanation (Dobyns 1992; Dobyns and Jahn 1995; Jefferys 1990, 1992, 1995). Many astronomers, physicists, chemists, and biologists believe it likely that the universe contains intelligent beings in a number of places. Many of them think it not unlikely that we could detect signals from such sources. Yet many other people would agree with the one who said, "If SETI [*S*earch for *E*xtra-*T*errestrial *I*ntelligence] does proceed, I suggest that we adopt the SCOTI Program . . . the search for congressional [terrestrial] intelligence" (Triplett 1992:80).

Some of those who accept that Fleischmann and Pons have achieved cold fusion of deuterium nuclei have gone further, asserting that other nuclear transformations also occur under similar conditions. Most people would respond that deuterium fusion is unlikely enough; for larger nuclei, how would they even get into the electrode lattice? Yet the believers can respond that, until we know the exact mechanism for the deuterium fusion, we have no right to exclude the possibility of others.

Anomalistics offers the opportunity to think hard about relative improbabilities and their possible significance. Ufology categorizes encounters as first kind—some sort of machine within five hundred feet or so; second kind—physical traces remain and can be analyzed; third kind—*beings* are encountered; fourth kind—humans are abducted. These are commonly taken to be in order of increasing improbability; yet, just as with *psi,* if all four are part of the same actual phenomenon, then their actual (im)probabilities are in fact identically the same.

Nor can one invoke the huge improbability of the fourth-kind encounters to reject the others; that one anomalistic claim is wrong does not disprove another apparently related one.

It is virtually certain that the oceans harbor yet-unidentified species. It is quite likely that some of those are fairly large. It is surely not impossible that one or other of those at some time became landlocked in what have become freshwater lakes. It is rather unlikely, though, that such a species—if it exists—is related most closely to some species thought extinct for tens of millions of years; it is even less likely that it's so different from any other extant species as a dinosaur or plesiosaur would be. So Nessies are indeed improbable. How much more improbable are they because some sightings have been reported *on land?*

Perhaps the various levels of improbability in anomalistics bear an analogy with the different orders of infinity identified by Georg Cantor (1845–1918): the infinity of points on a line is of a greater order than that of the prime numbers, for example. Three different orders of improbability in anomalistics might be envisaged, using the classification of what we know, what we *know* we don't know (the known unknown, say, the details of the genetic code), and what we don't even suspect is there to be known (the unknown unknown):

1. implausible by what we now believe based on sound experience (for instance, perpetual motion);
2. implausible even by what we can only conceive but believe to be so (for instance, inexplicable even by something like harnessing zero-point quantum fluctuations of the vacuum);
3. inconceivable: "science-fiction" notions, fantasy; say, the existence of "other dimensions" or "other forces" or "other frequencies," that might do for electromagnetics what relativity did for gravity and mechanics.

If nothing else, this classification reinforces the grounds for humility as to what in the future might emerge from the unknown unknown. Consider into which of these three orders of implausibility wise people might have placed:

- Before 1000 BCE: on the one hand, electricity, firearms, airplanes, television; on the other hand, extrasensory perception, faith healing, prophesy, psychokinesis

- Before 1900 CE, superconductivity (discovered 1911); quantum-mechanical tunneling; genetic engineering; desktop computers capable of umpteen MIPS
- Before 1975 CE, high-temperature superconductivity; bucky-balls (buckminsterfullerenes)

To Err Is Human — I. J. Good contends, "Many fallacies . . . have their origin in wishful thinking, laziness, and busyness. These . . . lead to oversimplification, the desire to win an argument at all costs (even at the cost of over*complication*), failure to listen to the opposition, too-ready acceptance of authority, too-ready rejection of it, too-ready acceptance of the printed word (even in newspapers), too-great reliance on a formal machine or formal system or formula (*deus ex machina*) . . . special pleading, the use of language in more than one sense without notice of the ambiguity (if the argument leads to a desirable conclusion), the insistence that a method used successfully in one field of research is the only appropriate one in another, the distortion of judgment, and the forgetting of the need for judgment" (Good 1978:337).

All those are lapses to which we are all prone, against which we need to guard all the time; and still, sooner or later we are likely to go wrong about something. The controversies about memories recovered under hypnosis, for instance, teach at the very least that absolute conviction that a memory is genuine is no proof that it is indeed so. It might be, but then again it might not be. We can each of us be wrong even when most sure that we are right.

What might make for competence at a task like anomalistics cannot be specified because the task itself cannot be well enough defined. The best cast of mind is one able to scan a situation with a minimum of preconceptions and to be incisively logical about the seemingly inexplicable. This is the quality possessed by G. K. Chesterton's Father Brown, for example, and approached perhaps by the best minds trained to the law or to criminal detection. There are many puzzling stories about absurdly impossible or inexplicable events that are seen to have perfectly rational explanations once the punch line is reached. A friend recently sent me these examples:

It's a fine sunny day in the forest, and a rabbit is sitting outside his burrow, tapping on his typewriter. Along comes a fox, out for a walk.

Fox: "What are you working on?"
Rabbit: "My thesis."
Fox: "Hmm. What's it about?"

Rabbit: "Oh, I'm writing about how rabbits eat foxes."
(incredulous pause)
Fox: "That's ridiculous! Any fool knows that rabbits don't eat foxes!"
Rabbit: "Come with me and I'll show you!"

They both disappear into the rabbit's burrow. After a few minutes, gnawing on a fox bone, the rabbit returns to his typewriter and resumes typing. Soon a wolf comes along and stops to watch the hardworking rabbit.

Wolf: "What's that you're writing?"
Rabbit: "I'm doing a thesis on how rabbits eat wolves."
(loud guffaws)
Wolf: "You don't expect to get such rubbish published, do you?"
Rabbit: "No problem. Want to see why?"

The rabbit and the wolf go into the burrow, and again the rabbit returns by himself, after a few minutes, and goes back to typing. Finally a bear comes along and asks, "What are you doing?"

Rabbit: "I'm doing a thesis on how rabbits eat bears."
Bear: "Well that's absurd!"
Rabbit: "Come into my home and I'll show you."

As they enter the burrow, the rabbit introduces the bear to the lion.

It's a fine sunny day in the forest. A lion is lying lazily in the sun outside his cave. Along comes a fox.

Fox: "Do you know the time? My watch is broken."
Lion: "Oh, I can easily fix the watch for you."
Fox: "Hmm. But it's a very complicated mechanism, and your great claws will only destroy it even more."
Lion: "Oh no, give it to me, you'll see, it will be fixed."
(incredulous pause)
Fox: "That's ridiculous! Any fool knows that lazy lions with great claws can't fix complicated watches."
Lion: "Sure they do, let me have it and you'll see."

The lion disappears into his cave, and after a while comes back with the watch, which is running perfectly. The fox is impressed, and the lion continues to lie lazily in the sun, looking very pleased with himself. Soon a wolf comes along.

Wolf: "Can I come and watch TV tonight with you? Mine's broken?"

Lion: "Oh, I can easily fix your TV for you."
(loud guffaws)
Wolf: "You don't expect me to believe such rubbish, do you? There's no way that a lazy lion with big claws can fix a complicated TV!"
Lion: "No problem. Do you want to try?"

The lion goes into his cave, and after a while comes back with a perfectly fixed TV. The wolf goes away happy and amazed.

Later, inside the lion's cave: In one corner are half-a-dozen small, intelligent-looking rabbits busily doing very complicated work with very detailed instruments. In the other corner lies a huge lion looking very pleased with himself.

Benefits of Anomalistics

The debunkers like to harp on the harm that is supposedly incurred by individuals and by society as a result of superstitious or pseudoscientific beliefs. To the contrary, as I've argued, there is no evidence that belief in the benefits of large doses of vitamin C or in personal auras is necessarily harmful to those who hold them, nor to those around those who hold them. Indeed, some of us have derived benefits from anomalist pursuits.

It's been suggested that we *need* mysteries: "Nowadays [1891!] we have so few mysteries left to us that we cannot afford to part with one of them" (Wilde 1891). Certainly many people like to hear about strange things. Many find it interesting to seek answers to what seem to be mysteries. Disbelievers will sometimes ask why people are attracted to such matters as ESP or UFOs when real science offers an endless array of wonderful mysteries to explore. One answer might be that the mysteries science grapples with are so specialized and technical as to seem quite closed to amateurs; they are not really accessible without long years of hard study, whereas anyone can become a UFO researcher or cryptozoologist almost forthwith. Thus anomalistics offers entrée to intellectual activity. Some anomalists have markedly developed intellectually through their anomalist quest, just as some people mature through mainstream scholarship or scientific research. The widespread interest aroused by scientific anomalies can also be put to use in the teaching of science (Howes and Watson 1981; Martin 1971).

Through investigating anomalies, we can learn about everyday matters that are more-or-less ignored in mainstream education or research. It was through a physiologist who wished to debunk UFOs that I (and many others at the

University of Sydney forty years ago) learned that we can see cells moving around on the surface of our corneas, if only we allow ourselves to notice them. Trying to make sense of anomalistic claims affords the opportunity to learn about a host of things like that, and about animal behavior, and a host of other natural phenomena.

There may also be quite tangible spinoff benefits for science and society from anomalistic investigations. At Loch Ness, a major seiche—an oscillation of water beneath the surface—recognized during sonar searches for monsters proved well worth studying; a population of arctic char was discovered; a World War II bomber was found and recovered and restored for exhibition in a museum.

Science as a whole can benefit from serious anomalistics, if only through making clearer in which areas our knowledge remains notably incomplete. William Corliss (1994) has gleaned from predominantly *scientific* periodicals a wealth of mysterious, unexplained things. The desire to assess the probable origin of materials purportedly dropped by UFOs revealed our lack of comprehensive knowledge of metallurgical capabilities coupled with the effects of ablation by heating as an object falls through the air. Gauquelin's Mars Effect led to years of discussion from which one could learn at the very least that it is no easy matter to select proper controls when looking for statistical correlations in human affairs. History, philosophy, and sociology of science are rather silent about what brings inspiration and spurs creativity. Anomalistic quests, we know as a matter of empirical fact, have sometimes served that purpose.

Scientists and students of science know the truth of the aphorism that it is less important for science to find answers than to find the right questions to ask. Anomalistics, through its persistent attention to what science does not yet know, offers a rich collection of intriguing questions. What science now regards as breakthroughs of the past were often, at the time, ignored or resisted (Barber 1961) or even dismissed as outrageous nonsense. "It is a classic dilemma for the scientist. . . . 'To make . . . progress, violence must be done to many of our accepted principles'" (Baker 1978:1255, quoting Davis 1926:464), such as, in the 1920s, "seriously consider[ing] 'the Wegener outrage of wandering continents.' . . . [We need] not . . . 'merely a brief contemplation followed by an offhand verdict of "impossible" or "absurd,"' but a 'contemplation deliberate enough to seek out just what conditions would make the outrage seem permissible and reasonable'" (Baker 1978:1256, quoting Davis 1926:464, 467–68).

For science globally rather than in specific detail, anomalistics can be a needed balance to the hubris to which science and scientists are tempted in the wake of science's magnificent progress over the last few centuries and the high societal status science has thereby gained. Anomalistics is not science, but an appreciation of the difference between them lends deeper insight into what science actually is; after all, as in Kipling's "The English Flag," "What should they know of England who only England know?"

A conference about the Sasquatch (Bigfoot) provided these pertinent conclusions (Ames 1980:302):

> One . . . of the more enlightening discussions during the conference concerned the role of evidence and the nature of the scientific method. . . . What are the proper relations between the scientific establishment and those amateurs, like the Sasquatch investigators, who toil on the frontiers of knowledge? How can members of the scholarly establishment resolve the contradiction between their commitment to the fundamental importance of free inquiry and the intellectual conservatism that derives from their scientific skepticism? To what extent should a scholar risk his or her professional reputation by pursuing nonrespectable topics? Are amateur investigators denied research support because they lack professional credentials or because they . . . pursue illicit topics? . . .
>
> The conference also engendered a greater respect for the role of amateur investigators, those individuals who, for whatever reasons, felt impelled to go beyond the boundaries of established knowledge and to stretch their investigations beyond the limits of established scientific methods. It is not suggested here, however, that professional scientists and the amateur investigators should accept one another's standards and interests, as some advocated during the debates. To turn one perspective into the other, either by expanding the limits of science or by limiting the freedom of amateurs, would destroy the potentially creative contradictions that exist between the two. The urge to probe beyond the realm of established knowledge and certainty, to explore the anomalous and unknown worlds, and to criticize the scientific establishment for its self-interested motivation is no less important than the practice of science itself. At the same time, however, it is important to reaffirm the value of a scientific establishment that is conservative about the rule of evidence, respects its own theories, and pursues its own intellectual interests rather than those of outsiders. Both perspectives are needed, each doing what it alone can do best, though each should be continuously exposed to the other, with their contradictions expressed through creative criticism. The professionals and the amateurs can help to keep one another honest.

Solving puzzles can be satisfying. Anomalistics offers the most excruciatingly challenging puzzles. More standard puzzles—jigsaws, crosswords, and the like—are limited in their challenge by the fact that they are human constructions. It's known beforehand that they have an unambiguous answer and that one can, if need be, obtain that answer without solving the puzzle for oneself. Science offers less-standard puzzles; but tackling them is constrained by standard methods and knowledge and theories and by peer pressure, so that the puzzler's creativity and originality are constrained within rather strict bounds.

Those restraints are absent in anomalistics; only nature and the investigator's own insights, talents, imagination, and creativity limit how the quest is carried on. The mysteries, the puzzles, are truly marvelous ones, for the possible explanations are so peculiar, so weird; they contradict common sense, they're so improbable—and yet periodically one of these quests is vindicated, at least in part, and we're reminded once again that anything that *did* happen, *can* happen. Moreover the anomalistic quest is not narrowly focused, as is the scientific one, on natural phenomena. Ufology is not only about objects in the sky, it is also about human psychology, about folklore and mythology and history, about the possibility of extraterrestrial intelligences. Who could ask for more marvelous mysteries to explore?

Anomalist quests can also meet the high mythic standards of heroic epics. They can be quixotic, Arthurian quests for a Holy Grail, where much of the value of the quest lies in the honorable faith and motives and behavior of the seekers, not in any tangible success that might ensue—and so it is a faithful analog of human life.

Anomalistics, I believe, can be rewarding not only to those who engage in the quests but also to scholars of science and society. History of science, philosophy of science, and sociology of science have concerned themselves so far with quite limited parts of science, predominantly physics and astronomy and latterly genetics and evolution. Very little scholarly attention has been paid to scientific anomalies or heresies as such. The comparisons I was led to make while writing this book suggest that a concentrated examination of intellectual heresies can be not only fascinating in itself but could also enrich our understanding of mainstream science and scholarship.

References

Abrahams, Marc. 1992. "Newsmaker Interview: Martin Fleischmann, Cold Fusion Pioneer." *Journal of Irreproducible Results* 37, no. 4 (July–August): 8.

Academy for New Energy. 1997. Flyer for Fourth International Symposium on New Energy, mailed March.

Aïssa, J., P. Jurgens, W. Hsueh, and J. Benveniste. 1997. "Trans-Atlantic Transfer of Digitized Antigen Signal by Telephone Link." Paper presented at Congress of American Association of Immunologists, February, San Francisco.

Alcock, James E. 1981. *Parapsychology: Science or Magic?* Oxford: Pergamon Press.

———. 1990. *Science and Supernature.* Amherst, N.Y.: Prometheus.

———. 1996. "Extrasensory Perception." In *Encyclopedia of the Paranormal,* ed. Gordon Stein, 241–54. Amherst, N.Y.: Prometheus.

Allen, John Eliot, and Marjorie Burns, with Samuel C. Sargent. 1986. *Cataclysms on the Columbia: A Layman's Guide to the Features Produced by the Catastrophic Bretz Floods in the Pacific Northwest.* Portland, Ore.: Timber Press.

Allen-Mills, Tony. 1998. "All Set for the Smithsonian of Sex." *Sunday Times* (London), 5 July, World News sec., p. 25.

Amato, Ivan. 1991. "Briefings: Genes Score a New Point in Alcoholism." *Science* 253 (26 July): 379.

———. 1993. "Random Samples—Closing the Case on Cluster Impact Fusion." *Science* 262 (22 October): 509.

Ames, Michael M. 1980. "Epilogue." In *Manlike Monsters on Trial: Early Records and Modern Evidence,* ed. Marjorie Halpin and Michael M. Ames, 301–15. Vancouver: University of British Columbia Press.

Andreski, Stanislav. 1972. *Social Sciences as Sorcery.* London: Andre Deutsch.

Andrews, Edmund. 1992. "Electromagnets to Help Plants Grow." *San Francisco Chronicle,* 11 January, p. A19, citing patent 5,077,934.

Anon. 1975. "Naming the Loch Ness Monster." *Nature* 258 (11 December): 466–68.

———. 1980. "Electric Charges May Shape Living Tissue." *New Scientist* 86:245.

———. 1986. "Gordon Research Conferences Announced." *Science* 231 (7 March): 1166.

———. 1989a. "Yellow Rain Is Bee Waste, Not Soviet Weapon, Scientists Say." *Columbus Dispatch,* 21 September, p. 6A.

———. 1989b. "Tell the Truth about Yellow Rain" (editorial). *St. Louis Post-Dispatch,* 4 October.

———. 1993. "In Stormy Days, Strokes Run Higher." *Roanoke Times,* 14 February, p. A13.

———. 1994a. "Fake in the Lake." *People,* 28 March, p. 109.

———. 1994b. "Cold Fusion Reproduced—on Paper." *Science* 264 (6 May): 771.

———. 1996a. "Blood Surge during Stress Harmful?" *Roanoke Times,* 31 January, p. A7.

———. 1996b. "Study: Gene Doesn't Influence Actions." *Roanoke Times,* 2 November, p. A5.

———. 1997a. "Gem Warfare." *Sunday Times* (London), 31 August, Style sec., p. 35.

———. 1997b. "Medicine Man, M.D.—Dr. Lewis Mehl-Madrona." *Whole Life Times—The Journal of Holistic Living,* October, pp. 49, 77.

———. 1997c. "Panel to Monitor Gulf War Illness Research." *Roanoke Times,* 9 November, p. A18.

———. 1997d. "Editors' Choice." *New York Times Book Review,* 7 December, p. 12.

———. 1998a. *Chronicle of Higher Education,* 9 January, p. A33.

———. 1998b. "S&M at SUNY." *CAMPUS* 9, no. 2 (Winter): 9.

———. 1998c. "Ancient Skull Adds to Mystery of Earliest Americans' Origins." *Roanoke Times,* 22 May, p. A6.

———. 1998d. Advertisement for *The Erotic Review. Sunday Times* (London), 5 July, World News, p. 1-24.

Arndts, Russell. 1989. "Cold Nuclear Fusion: Its Implications for Science and the Origins Debate." *Bible-Science Newsletter,* June, p. 11.

Arnold, F. W. 1981. "Cell Control." *New Scientist,* 19 March, p. 766.

Arp, Halton. 1987. *Quasars, Redshifts and Controversies.* Berkeley, Calif.: Interstellar Media.

———. 1998. *Seeing Red—Redshifts, Cosmology and Academic Science.* Montreal: Apeiron.

Ashburner, M. 1981a. "Mysterious Dr X." *New Scientist,* 12 February, p. 433.

———. 1981b. "Credibility." *New Scientist,* 7 May, p. 378.

Ashford, Nicholas A., and Claudia S. Miller. 1990. "Chemical Sensitivity." *Chemical and Engineering News,* 12 March, p. 2.

Ashworth, William. 1980. "Sharks' Teeth, Axe-Heads and Belemnites: Problems of Nomenclature in Early Seventeenth-Century Paleontology." Paper presented at "Science and Technology Studies: Toronto 80," joint meeting of the History of Science Society, Philosophy of Science Association, Society for the History of Technology, and Society for the Social Study of Science, October, Toronto, Canada.

Atkins, P. W., and T. P. Lambert. 1975. "The Effect of a Magnetic Field on Chemical Reactions." *Annual Reports on the Progress of Chemistry,* section A, 72:67–88.

Atkins, Peter. 1976. "Magnetic Field Effects." *Chemistry in Britain* 12:214–18, 228.

Aveni, Anthony F. 1989. *World Archaeoastronomy.* Cambridge: Cambridge University Press.

———. 1992. *Conversing with the Planets: How Science and Myth Invented the Cosmos.* New York: Times Books.

———. 1996. *Behind the Crystal Ball: Magic, Science, and the Occult from Antiquity through the New Age.* New York: Random House.

Awareness (magazine). 1997. September–October.

Baity, Elizabeth Chesley. 1973. "Archaeoastronomy and Ethnoastronomy So Far." *Current Anthropology* 14 (October): 389–449.

Baker, R. Robin. 1980. "Goal Orientation by Blindfolded Humans after Long-Distance Displacement: Possible Involvement of a Magnetic Sense." *Science* 210 (31 October): 555–57.

———. 1989. *Human Navigation and Magnetoreception.* Manchester, England: Manchester University Press.

Baker, Victor R. 1978. "The Spokane Flood Controversy and the Martian Outflow Channels." *Science* 202 (22 December): 1249–56.

Barber, Bernard. 1961. "Resistance by Scientists to Scientific Discovery." *Science* 134 (1 September): 596–602.

Barron, Laurence D. 1994. "Can a Magnetic Field Induce Absolute Asymmetric Synthesis?" *Science* 266 (2 December): 1491–92.

Barry, James Dale. 1980. *Ball Lightning and Bead Lightning.* New York: Plenum.

Barzun, Jacques. 1964. *Science: The Glorious Entertainment.* New York: Harper and Row.

Bassett, C. Andrew L. 1995. "Bioelectromagnetics in the Service of Medicine." In *Electromagnetic Fields—Biological Interactions and Mechanisms,* ed. Martin Blank, 261–75. Advances in Chemistry, no. 250. Washington, D.C.: American Chemical Society.

Bauer, Henry H. 1972. *Electrodics—Modern Ideas Concerning Electrode Reactions.* Stuttgart: Georg Thieme.

———. 1982. "Comments to Ron Westrum's 'Crypto-Science and Social Intelligence about Anomalies.'" *Zetetic Scholar* 10 (December): 109–11.

———. 1983. "Velikovsky and the Loch Ness Monster: Attempts at Demarcation in Two Controversies." In "Working Papers of the Center for the Study of Science in Society (VPI&SU)," ed. Rachel Laudan, 2, no. 1 (April): 87–106.

———. 1984a. *Beyond Velikovsky: The History of a Public Controversy.* Urbana: University of Illinois Press.

———. 1984b. "Velikovsky and Social Studies of Science." *4S Review* 2, no. 4 (Winter): 2–8.

———. 1986. *The Enigma of Loch Ness: Making Sense of a Mystery.* Urbana: University of Illinois Press.

———. 1986–87. "The Literature of Fringe Science." *Skeptical Inquirer* 11, no. 2 (Winter): 205–10.

———. 1987. "Distributions of Belief on Controversial Matters." *Zetetic Scholar* 12–13 (August): 99–102.

———. 1990a. "Barriers against Interdisciplinarity: Implications for Studies of Science, Technology, and Society (STS)." *Science, Technology, and Human Values* 15:105–19.

———. 1990b. "A Dialectical Discussion on the Nature of Disciplines and Disciplinarity." *Social Epistemology* 4:215–27.

———. 1991. *Journal of Scientific Exploration* 5:267–70; review of Close (1991).

———. 1992a. *Scientific Literacy and the Myth of the Scientific Method.* Urbana: University of Illinois Press.

———. 1992b. *Journal of Scientific Exploration* 6:395–400; review of Huizenga (1992).

———. 1992c. *Journal of Scientific Exploration* 6:81–84; review of Mallove (1991).

———. 1993–96. *Cryptozoology* 12:80–82; review of Shiels (1990).

———. 1995a. "Sexpols, Orgone Energy, Cloud Busters, and Bions: Remembered, but Not Relevant." *Skeptical Inquirer* 19, no. 5 (September–October): 43–45; review of Higgins (1994).

———. 1995b. "Two Kinds of Knowledge: Maps and Stories." *Journal of Scientific Exploration* 9:257–75.

———. 1995c. *Journal of Scientific Exploration* 9:301–7; review of Taubes (1993).

———. 1997. "A Consumer's Guide to Science Punditry." In *Science Today: Problem or Crisis?* ed. Ralph Levinson and Jeff Thomas, 22–34. London: Routledge.

Bauer, Henry H., Gary D. Christian, and James E. O'Reilly. 1978. *Instrumental Analysis.* Boston: Allyn and Bacon.

Bauer, Henry H., and Patrick Huyghe. 1999. "Fatal Attractions: The Troubles with Science." Manuscript.

Baum, Rudy. 1992. "Biological Magnets Found in Human Brain." *Chemical and Engineering News,* 18 May, p. 6.

Bayanov, Dmitri. 1982. "A Note on Folklore in Hominology." *Cryptozoology* 1:46–48.

BBC Television. 1994. *Heretics.* A series of six half-hour documentaries shown on BBC Television, summer.

Beaudette, Charles G. "Cold Fusion and the Press." 1993. *The Torch* 66, 1 (Fall).

Becker, Robert O., and Gary Selden. 1985. *The Body Electric: Electromagnetism and the Foundation of Life.* New York: William Morrow.

Beckjord, Erik. 1988. Quoted in Helen Jacks, "Time for New Nessie Approach—Expert," *Aberdeen Press and Journal* (Scotland), 25 August.

Begley, Sharon, with Steve Levin. 1994. "All that Glitters Isn't Chemistry." *Newsweek,* 10 January, p. 54.

Behe, Michael J. 1996. *Darwin's Black Box: The Biochemical Challenge to Evolution.* New York: Free Press.

Beloff, John. 1993. *Parapsychology: A Concise History.* New York: St. Martin's.

Beloff, John, with James Alcock, Irvin L. Child, Daniel Cohen, H. M. Collins, Robert L. Morris, J. Ricardo Musso, Mirta Granero, J. Fraser Nicol, John Palmer, K. Ramakrishna Rao, James Randi, Christopher Scott, Sybo Schouten, and Rex G. Stanford. 1980. "Seven Evidential Experiments." *Zetetic Scholar* 6 (July): 91–120.

Bem, Daryl J., and Charles Honorton. 1994. "Does Psi Exist? Replicable Evidence for an Anomalous Process of Information Transfer." *Psychological Bulletin* 115, no. 1 (January): 4–18.

Benjamin, Jonathan, Lin Li, Chavis Patterson, Benjamin D. Greenberg, Dennis L. Murphy, and Dean J. Hamer. 1996. "Population and Familial Association between the D4 Dopamine Receptor Gene and Measures of Novelty Seeking." *Nature Genetics* 12 (January): 81–84.

Bennion, Bruce C., and Laurence A. Neuton. 1976. "The Epidemiology of Research on 'Anomalous Water.'" *Journal of the American Society for Information Science* (January–February): 53–56.

Berger, Arthur S., and Joyce Berger. 1991. *The Encyclopedia of Parapsychology and Psychical Research.* St. Paul, Minn.: Paragon House.

Bernstein, Jeremy. 1967. *A Comprehensible World.* New York: Random House.

———. 1982. *Science Observed.* New York: Basic Books.

Berrettini, Wade H., Thomas N. Ferraro, Lynn R. Goldin, Daniel E. Weeks, Sevilla Detera-Wadleigh, John I. Nurmberger Jr., and Elliot S. Gershon. 1994. *Proceedings of the National Academy U.S.A.* 91:5918–21.

Binns, Ronald. 1984. *The Loch Ness Mystery Solved.* Buffalo, N.Y.: Prometheus.

Bio-Electro-Magnetics Institute. 1989. *BEMI Currents* (Newsletter) 1, no. 1 (Spring).

Birgham, Francis. 1881. "The Discovery of Organic Remains in Meteoritic Stones." *Popular Science Monthly* 20:83.

Blank, Martin, ed. 1995. *Electromagnetic Fields—Biological Interactions and Mechanisms.* Advances in Chemistry, no. 250. Washington, D.C.: American Chemical Society.

Boadella, David. 1974. *Wilhelm Reich: The Evolution of His Work.* Chicago: Henry Regnery.

Bonk, Thomas. 1997. "Swede Dreams at Troon." *Roanoke Times,* 20 July, p. B1.

Borgens, Richard B., Ernesto Roederer, and Melvin J. Cohen. 1981. "Enhanced Spinal Cord Regeneration in Lamprey by Applied Electric Fields." *Science* 213 (7 August): 611–17.

Boyd, Alastair, and David Martin. 1994. "Creating a Monster." *BBC Wildlife,* April, pp. 22–23.

Bradley, Richard. 1997. "What's New at the Clava Cairns?" Public lecture, 24 August, Culloden Centre, Inverness, Scotland.

Breyer, B., and Henry H. Bauer. 1963. *Alternating Current Polarography and Tensammetry.* New York: Interscience.

Bright-Life. 1997. Catalog BRT119-CAT/7.97, p. 14. Hicksville, N.Y.

Britz, Dieter. 1990. "Cold Fusion: An Historical Parallel." *Centaurus* 33:368–72.

Broad, William, and Nicholas Wade. 1982. *Betrayers of the Truth: Fraud and Deceit in the Halls of Science.* New York: Simon and Schuster.

Brodeur, Paul. 1989. *Currents of Death: Power Lines, Computer Terminals, and the Attempt to Cover Up Their Threat to Your Health.* New York: Simon and Schuster.

Bronowski, Jacob. 1973. *The Ascent of Man.* Boston: Little, Brown.

Brooks, David. 1998. "Econ-Icons." *New York Times Book Review,* 31 May, p. 54; review of *Two Lucky People,* by Milton Friedman and Rose D. Friedman.

Broughton, Richard S. 1992. *Parapsychology: The Controversial Science.* New York: Ballantine.

Brown, David. 1997. "Critics Cry Foul over $3 Million Study Funded outside Usual Channels." *Roanoke Times,* 9 November, p. A18.

Brush, Stephen. 1978. *The Temperature of History.* New York: Burt Franklin.

Bryant, Clifton D. 1982. *Sexual Deviancy and Social Proscription.* New York: Human Sciences Press.

Bullard, Thomas E. 1988. "The Mechanization of UFOs." *International UFO Reporter* 13, no. 1 (January–February): 8–12.

Burda, Gwen A. 1997. "The Battle between Political Agendas and Science." *Skeptical Inquirer* 21, no. 6 (November–December): 13–15.

Burgstahler, Albert W. 1988. "Fluoridation of Water." *Chemical and Engineering News,* 17 October, p. 3.

Burr, Harold Saxon. 1972. *Blueprint for Immortality: The Electric Patterns of Life.* London: Neville Spearman.

Burton, Maurice. 1961. *The Elusive Monster.* London: Rupert Hart-Davis.

Byrne, Gregory. 1988. "Breuning Pleads Guilty." *Science* 242 (7 October): 27–28.

Cairns, John, Jr. 1997. Personal communication, December.

Cambridge Loch Ness Expedition. 1962. Report.

Campbell, Steuart. 1991. *The Loch Ness Monster: The Evidence.* Old Tappan, N.J.: Macmillan.

———. 1995. "Nessie 'Model' Explanation Suspect." *Skeptical Inquirer* 19, no. 2 (March–April): 62–63.

Carlinsky, Joel. 1990. "Orgonomy in New Jersey." *New York Skeptic* 3, no. 2 (Fall).

———. 1994. "Epigones of Orgonomy—the Incredible History of Wilhelm Reich and His Followers." *Skeptic* 2 (3): 90–92.

———. 1996. "Orgonomy Peddlers." *Paranoia* 13.

———. 1998. Personal communication.

———. n.d. "(Or)gone but Not Forgotten: The Legacy of Wilhelm Reich Lives on in His Modern Followers." *KASES File* (newsletter of the Kentucky Skeptics group), pp. 6–7.

Carter, Harriet. n.d. Mail-order catalog K37–9, received 1998, pp. 3, 4, 25, 27, 78, 85, 86, 112, 115.

Cavendish, Richard, ed. 1989. *Encyclopedia of the Unexplained—Magic, Occultism and Parapsychology.* London: Arkana.

Center for Inquiry. 1996. Eight-page brochure for "Fund for the Future" with covering letter, 19 September. Amherst, N.Y.

———. 1998. "Winter Session" flyer.

Cerf, Christopher, and Victor Navasky. 1984. *The Experts Speak: The Definitive Compendium of Authoritative Misinformation.* New York: Pantheon.

Chambers, Whittaker. 1953. *Witness.* London: Andre Deutsch.

Chapman, David. 1991. Sci.physics.fusion newsgroup, 5 June.

Christman, David R. 1990. "Cold Fusion." *Chemical and Engineering News,* 17 September, p. 78.

Clark, Jerome. 1987. "Ness Sense and Nonsense." *Fate* 40 (February): 99–105.
———. 1993. *Encyclopedia of Strange and Unexplained Physical Phenomena.* Detroit, Mich.: Gale Research.
———. 1998. *The UFO Encyclopedia.* 2d ed. Detroit, Mich.: Omnigraphics.
Clarke, Arthur C. 1980. Introduction. In *Arthur C. Clarke's Mysterious World,* ed. Simon Welfare and John Fairley. London: Collins.
———. 1984a. Introduction. In *Arthur C. Clarke's World of Strange Powers,* ed. John Fairley and Simon Welfare. New York: G. B. Putnam's Sons.
———. 1984b. Epilogue. In *Arthur C. Clarke's World of Strange Powers,* ed. John Fairley and Simon Welfare. New York: G. B. Putnam's Sons.
Close, Frank. 1991. *Too Hot to Handle: The Race for Cold Fusion.* Princeton: Princeton University Press.
Clube, Victor, and Bill Napier. 1982. *The Cosmic Serpent: A Catastrophist View of Earth History.* New York: Universe Books.
Clustron Sciences Corporation. 1992. News release, Vienna, Va., 10 August.
Coe, Michael D. 1992. *Breaking the Maya Code.* New York: Thames and Hudson.
Coggins, Catherine. 1976. "Loch Ness Monster: Fact or Fable? Question Divides Scientific Community." *Probe the Unknown,* May, pp. 33–39, 58–60.
Cohen, Daniel. 1965. *Myths of the Space Age.* New York: Dodd, Mead.
Coly, Lisette, and Joanne D. McMahon, eds. 1993a. *PSI Research Methodology: A Re-Examination.* New York: Parapsychology Foundation.
———. 1993b. *PSI and Clinical Practice.* New York: Parapsychology Foundation.
Corliss, William R. 1994. *Science Frontiers: Some Anomalies and Curiosities of Nature.* Glen Arm, Md.: The Sourcebook Project. First eighty-six issues, 1976–1993, of bimonthly newsletter *Science Frontiers.*
Coughlin, Ellen K. 1988. "Scholar Who Submitted Bogus Article to Journals May Be Disciplined." *Chronicle of Higher Education,* 2 November, pp. A1, 7.
———. 1995. "Psychotherapy Besieged: Research Psychologists and Insurers Question the Scientific Validity of Clinical Practices." *Chronicle of Higher Education,* 4 August, pp. A7, 12.
Craig, Gordon A. 1990. "Be Quiet, Father Is Writing." *New York Times Book Review,* 16 September, pp. 15–16.
Crease, Robert P. 1989. "The Epilepsy 'Cure': Bold Claims, Weak Data." *Science* 245 (29 September): 1444–45.
Crick, Francis. 1981. *Life Itself.* New York: Simon and Schuster.
Crist, Peter A. 1995. "The American College of Orgonomy Today," *A.C.O. Newsletter* (Summer).
———. 1997. "Peter Reich to Keynote A.C.O.'s Wilhelm Reich Centenary Celebration." *A.C.O. Newsletter* (Special Edition).
Croswell, Ken. 1997. *Planet Quest: The Epic Discovery of Alien Solar Systems.* New York: Free Press.
Dagani, Ron. 1989. "Nuclear Fusion: Utah Findings Raise Hopes, Doubts." *Chemical and Engineering News,* 3 April, pp. 4–6.

———. 1993. "Latest Cold Fusion Results Fail to Win Over Skeptics." *Chemical an Engineering News,* 14 June, pp. 38–41.

———. 1995a. "Latest Research on Cold Fusion to be Presented at Anaheim AC Meeting." *Chemical and Engineering News,* 20 March, pp. 22–23.

———. 1995b. "Cold Fusion Session Smaller Than Expected." *Chemical and Engineer ing News,* 10 April, p. 7.

———. 1996. "Cold Fusion Lives—Sort Of." *Chemical and Engineering News,* 29 April p. 69.

Davenas, E., F. Beauvais, J. Amara, M. Oberbaum, B. Robinzon, A. Miadonna, A. Te deschi, B. Pomeranz, P. Fortner, P. Belon, J. Sainte-Laudy, B. Poitevin, and J. Ben veniste. 1988. "Human Basophil Degranulation Triggered by Very Dilute Antiserun Against IgE." *Nature* 333 (30 June): 816–18.

Davis, W. M. 1926. "The Value of Outrageous Geological Hypotheses." *Science* 43 (May): 463–68.

De Beauregard, Olivier Costa, Richard D. Mattuck, Brian D. Josephson, and Evan Harri Walker. 1980. Letter to the Editor. *New York Review of Books,* 26 June.

Debus, Allen G. 1978. "Science vs. Pseudo-Science: The Persistent Debate." Inaugura Lecture, Morris Fishbein Center for the Study of the History of Science and Medicine, 27 November (published 1979).

De Camp, L. Sprague. 1976. "Little Green Men from Afar." *Humanist* 36, no. 1 (July–August): 5–8.

De Maigret, Pamela. 1994. *Journal of Scientific Exploration* 8:292–94: review of *Ambros Bierce Is Missing and Other Historical Mysteries,* by Joe Nickell.

DeMeo, James. 1989. "Response to Martin Gardner's Attack on Reich and Orgon Research in the *Skeptical Inquirer.*" *Pulse of the Planet* 1, no. 1 (Spring): 11–17.

———. 1992. December mailing to "Dear Friend of the Life Energy."

———. 1994. "In Defense of Orgonomy." *Skeptic* 2 (no. 4): 15–17.

Derjaguin, B. 1983. "Polywater Reviewed." *Nature* 301 (6 January): 9–10.

Derra, Skip. 1989. "Will Science Ever Recover from the Cold Fusion Furor?" *Research and Development,* July, pp. 33, 34, 39.

De Russy, Candace. 1998. "'Revolting Behavior': The Irresponsible Exercise of Academic Freedom." *Chronicle of Higher Education,* 6 March, p. B9.

Devereux, Paul. 1991. "Three-Dimensional Aspects of Apparent Relationships between Selected Natural and Artificial Features within the Topography of the Avebury Complex." *Antiquity* 65:894–98.

Dixon, E. James. 1993. *Quest for the Origins of the First Americans.* Albuquerque: University of New Mexico Press.

Dobyns, York H. 1992. "On the Bayesian Analysis of REG Data." *Journal of Scientific Exploration* 6:23–45.

Dobyns, York H., and Robert G. Jahn. 1995. "Response to Jefferys." *Journal of Scientific Exploration* 9:122–24.

Donovan, Stephen K., ed. 1989. *Mass Extinctions: Processes and Evidence.* New York: Columbia University Press.

Dotto, Lydia. 1978. "Science Confronts 'Pseudo-Science.'" *Science Forum*, July–August, p. 21.
Drosnin, Michael. 1997. *The Bible Code.* New York: Simon and Schuster.
Duesberg, Peter. 1996. *Inventing the AIDS Virus.* Washington, D.C.: Regnery.
Duster, T. 1992. "Genetics, Race, and Crime: Recurring Seduction to a False Precision." In *DNA on Trial, Genetic Identification and Criminal Justice,* ed. P. Billings, 129–40. Cold Spring Harbor, N.Y.: Cold Spring Harbor Press.
Ebstein, Richard B., Olga Novick, Roberto Umansky, Beatrice Priel, Yamima Osher, Darren Blaine, Estelle R. Bennett, Lubov Nemanov, Miri Katz, and Robert H. Belmaker. 1996. "Dopamine D4 Receptor (*D4DR*) exon III Polymorphism Associated with the Human Personality Trait of Novelty Seeking." *Nature Genetics* 12 (January): 78–80.
Edge, D. O. 1973. "Technological Metaphor." In *Meaning and Control,* ed. D. O. Edge and J. N. Wolfe, 31–59. London: Tavistock.
Edwards, Paul. 1967. "Reich, Wilhelm." In *The Encyclopedia of Philosophy,* volume 7, 104–15. New York: Macmillan.
———. 1974. "The Greatness of Wilhelm Reich." *Humanist* 34 (March–April): 32–35.
Eisenberg, David. 1981. "A Scientific Gold Rush." *Science* 213 (4 September): 1104–5; review of Franks (1981).
Ellis, Albert. 1989. *Why Some Therapies Don't Work.* Buffalo, N.Y.: Prometheus.
Ellis, Albert, and Robert A. Harper. 1975. *A New Guide to Rational Living.* North Hollywood, Calif.: Wilshire Book Co.
Ember, Lois R. 1984. "Yellow Rain." *Chemical and Engineering News,* 9 January, pp. 8–34.
———. 1986. "New Data Weaken U.S. Yellow Rain Case." *Chemical and Engineering News,* 9 June, pp. 23–24.
Epstein, Michael. 1993. "The Skeptical Perspective." *Journal of Scientific Exploration* 7:15–18.
Estling, Ralph. 1982. "The Principle of Inverse Irreversibility." *New Scientist,* 23–30 December, pp. 808–10.
———. 1983. "The Trouble with Thinking Backwards." *New Scientist,* 2 June, pp. 619–21.
Evans, J. V., and G. N. Taylor. 1959. "Radio Echo Observations of Venus." *Nature* 184 (31 October): 1358.
Farley, John. 1977. *The Spontaneous Generation Controversy from Descartes to Oparin.* Baltimore: Johns Hopkins University Press.
Feder, Kenneth L. 1986. "The Challenges of Pseudoscience." *Journal of College Science Teaching* (December 1985–January): 180–86.
Felt, Ulrike, and Helga Nowotny. 1992. "Striking Gold in the 1990s: The Discovery of High-Temperature Superconductivity and Its Impact on the Science System." *Science, Technology and Human Values* 17:506–31.
Fischer, Steven Roger. 1997. *Glyphbreaker.* New York: Copernicus/Springer.
Flam, Faye. 1992a. "Survival of the Fittest in 1992's Physics and Astronomy Bestiary—A Quantized Stairway to Heaven?" *Science* 258 (18 December): 1884.

———. 1992b. "Survival of the Fittest in 1992's Physics and Astronomy Bestiary—Planets Come and Planets Go." *Science* 258 (18 December): 1885.

Fleischmann, M., and S. Pons. 1993. "Calorimetry of the Pd-D_2O System: From Simplicity via Complications to Simplicity." *Physics Letters A* 176:1.

Ford, Brian J. 1982. "Hermits, Crackpots and Pseudoscience." *Science Digest,* February, pp. 106–7.

Fort, Charles. 1974. *The Complete Books of Charles Fort.* New York: Dover.

Forum Parawissenschaften. 1999. <http://www.naa.net/forum_parawissenschaften>.

Forward, Robert L. 1996. "Mass Modification Experiment Definition Study (an Air Force Report)." *Journal of Scientific Exploration* 10:325–54.

Foster, Kenneth R. 1995. *IEEE Engineering in Medicine and Biology* 15, no. 6 (November–December 1996): 135; review of Malmivuo and Plonsey (1995).

Foulkes, R. A. 1971. "Dowsing Experiments." *Nature* 229 (15 January): 163–68.

Fraknoi, Andrew. 1984. "Scientific Responses to Pseudoscience." *Mercury* 13 (July–August): 121–24.

Frank, Louis A., with Patrick Huyghe. 1990. *The Big Splash.* New York: Birch Lane (Carol).

Franklin, Allan. 1993. *The Rise and Fall of the "Fifth Force": Discovery, Pursuit, and Justification in Modern Physics.* New York: Springer-Verlag.

Franks, Felix. 1981. *Polywater.* Cambridge, Mass.: MIT Press.

Frazier, Kendrick. 1976. "Science and the Parascience Cults." *Science News* 109 (29 May): 346–50.

———. 1978. "UFOs, Horoscopes, Bigfoot, Psychics and Other Nonsense." *Smithsonian* 8 (December): 55–60.

Freeman, Derek. 1983. *Margaret Mead and Samoa: The Making and Unmaking of an Anthropological Myth.* Cambridge, Mass.: Harvard University Press.

Friedman, William F., and Elizabeth S. Friedman. 1957. *The Shakespearean Ciphers Examined.* Cambridge: Cambridge University Press.

Fusion Facts. 1994. Vol. 6, no. 4 (October).

Gardner, Martin. 1950. "The Hermit Scientist." *Antioch Review* 10:447–57.

———. 1957. *Fads and Fallacies in the Name of Science.* New York: Dover. (First published as *In the Name of Science,* Putnam, 1952.)

———. 1985. "Welcome to the Debunking Club." *Skeptical Inquirer* 9, no. 4 (Summer): 319–22.

———. 1988. "Notes of a Fringe-Watcher—Reich the Rainmaker: The Orgone Obsession." *Skeptical Inquirer* 13, no. 1 (Fall): 26–30.

———. 1997. "Intelligent Design and Philip Johnson." *Skeptical Inquirer* 21, no. 6 (November–December): 17–20.

Geller, Robert C., David D. Jackson, Yan Y. Kagan, and Francesco Mulargia. 1997. "Earthquakes Cannot be Predicted." *Science* 275 (14 March): 1616–17.

Gellner, Ernest. 1984. "The Scientific Status of the Social Sciences." *International Social Science Journal* 36:567–86.

Gene Emergence Project. 2000. <http://www.us.net/life> (updated 1 February 2000).

George, Leonard. 1995. *Alternative Realities: The Paranormal, the Mystic and the Transcendent in Human Experience.* New York: Facts on File.

Gilovich, Thomas. 1991. *How We Know What Isn't So.* New York: Free Press.

Glanz, James. 1996a. "Year of Strange Events Leaves Standard Theory Unscathed." *Science* 274 (11 October): 179.

———. 1996b. "The Spell of Sonoluminescence." *Science* 274 (1 November): 718–19.

———. 1997. "Is First Extrasolar Planet a Lost World?" *Science* 275 (25 February): 1257–58.

Glut, Donald F. 1973. *The Frankenstein Legend: A Tribute to Mary Shelley and Boris Karloff.* Metuchen, N.J.: Scarecrow.

Gold, Thomas. 1999. *The Deep Hot Biosphere.* New York: Copernicus.

Good, Irving John. 1953. *Annals of Eugenics* 18:263–66; review of *Design and Analysis of Experiment,* by M. H. Quenoille.

———. 1962a. "A Classification of Fallacious Arguments and Interpretations." *Technometrics* 4, no. 1 (February): 125–32.

———. 1966. "The Future of Divination." *ARK—Journal of the Royal College of Art* 40 (Summer): 29–33.

———. 1974. "Random Thoughts about Randomness." In *PSA 1972,* Boston Studies in the Philosophy of Science, ed. K. F. Schaffner and R. S. Cohen, 117–35 (Dordrecht: Reidel); reprinted in *Good Thinking: The Foundations of Probability and Its Applications* (Minneapolis: University of Minnesota Press, 1983).

———. 1978. "Fallacies, Statistical." In *International Encyclopedia of Statistics,* ed. William H. Kruskal and Judith M. Tanur, vol. 1, 337–49. New York: Free Press.

———. 1980. "Scientific Speculations on the Paranormal and the Parasciences." *Zetetic Scholar* 7 (December): 9–29.

———. 1992. "The Bayes/Non-Bayes Compromise: A Brief Review." *Journal of the American Statistical Association* 87 (September): 597–602.

———. 1998. "The Self-Consistency of the Kinematics of Special Relativity, IV." *Physics Essays* 11 (no. 2): 248–63.

———, ed. 1962b. *The Scientist Speculates: An Anthology of Partly-Baked Ideas.* London: Heinemann.

Goodman, Eugene M., Ben Greenebaum, and Michael T. Marron. 1995. "Effects of Electromagnetic Fields on Molecules and Cells." *International Review of Cytology* 158:279–338.

Graneau, Peter. 1994. "Capillary Fusion." *Journal of Scientific Exploration* 8:428–29.

Gratzer, Walter. 1987. "A Buyer's Market." *Nature* 329 (3 September): 23; review of *Communicating Science to the Public,* ed. David Evered and Maeve O'Connor.

Gray, Thomas. 1984. "University Course Reduces Belief in Paranormal." *Skeptical Inquirer* 8, no. 3 (Spring): 247–51.

———. 1987. "Educational Experience and Belief in Paranormal Phenomena." In *Cult Archaeology and Creationism,* ed. Francis B. Harrold and Raymond A. Eve, 21–33. Iowa City: University of Iowa Press.

Grayson, Donald K. 1983. *The Establishment of Human Antiquity.* New York: Academic Press.

Greenberg, David. 1998. "Party Faithful." *New York Times Book Review,* 11 January, p. 22; review of *The Christian Coalition,* by Justin Watson.

Greenfield, Jerome. 1974. *Wilhelm Reich vs. the U.S.A.* New York: Norton.

Grinnell, Frederick. 1996. "Ambiguity in the Practice of Science." *Science* 272 (19 April): 333.

Gross, Paul R., and Norman Levitt. 1994. *Higher Superstition: The Academic Left and Its Quarrels with Science.* Baltimore: Johns Hopkins University Press.

Group for the Scientific Reappraisal of the HIV-AIDS Hypothesis. 1995. Letter. *Science* 267 (17 February): 945–46; also on the Internet (in 1998) at <http://www.virusmyth.com/aids>.

Gruenberger, Fred J. 1964. "A Measure for Crackpots." *Science* 145 (25 September): 1413–15.

Hall, R. N. 1968. *Proceedings of the Ninth International Conference on the Physics of Semiconductors.* Moscow.

Hamilton, Denise. 1997. "Suffering for Their Art." *New Times* (Los Angeles) 2, no. 10 (9–15 October): 11–15.

Handsome Rewards (19465 Brennan Ave., Perris, Calif.). 1996–97. Pamphlet F-43348R-1, p2599, item 10-51560-0.

Hannan, Patrick J., Rustum Roy, and John F. Christman. 1988. "Serendipity in Chemistry, Astronomy, Defense, and Other Useless Fields." *ChemTech,* July, pp. 402–6.

Hanneman, R. E., and P. J. Jorgensen. 1967. "On the Existence of Electromechanical and Photomechanical Effects in Semiconductors." *Journal of Applied Physics* 38:4099–4100.

Hansen, George P. 1992. "CSICOP and the Skeptics: An Overview." *Journal of the American Society for Psychical Research* 86:19–63.

Hanson, David. 1990. "Chemical Sensitivity: Growing Concern over Low Exposures." *Chemical and Engineering News,* 26 February, pp. 4–5.

Harrer, Heinrich. 1962. *The White Spider: The Story of the North Face of the Eiger.* Trans. Hugh Merrick. London: Rupert Hart-Davis.

Harrington, Anne, ed. 1997. *The Placebo Effect—An Interdisciplinary Exploration.* Cambridge, Mass.: Harvard University Press.

Harrison, Paul. 1999. *Encyclopedia of the Loch Ness Monster.* London: Robert Hale.

Hazen, Robert M. 1988. *The Breakthrough: The Race for the Superconductor.* New York: Summit Books.

Hetherington, Norris S. 1990. "Hubble's Cosmology." *American Scientist* 78 (March–April): 142–51.

Heuvelmans, Bernard. 1968. *In the Wake of the Sea-Serpents.* New York: Hill and Wang.

Hieronimus. 1997. *Hieronimus and Co. Newsletter* 2, no. 11.

Higgins, Mary Boyd, ed. 1994. *Beyond Psychology: Letters and Journals, 1934–1939, by Wilhelm Reich.* New York: Farrar, Straus, and Giroux.

Hileman, Bette. 1991. "Multiple Chemical Sensitivity." *Chemical and Engineering News,* 22 July, pp. 26–42.
Hillman, Harold. 1987a. "Oh to Be a Maverick." *New Scientist,* 1 October, p. 75.
———. 1987b. "The Price of Freedom." *New Scientist,* 22 October, p. 74.
———. 1988. "Hillman's Response." *The Scientist,* 19 September, pp. 14, 17.
Hodgkinson, Neville. 1993. "Nuclear Confusion." *Sunday Times* (London), 27 June, pp. 2, 6.
Hoffman, Nate. 1995. *A Dialogue on Chemically Induced Nuclear Effects: A Guide for the Perplexed about Cold Fusion.* La Grange Park, Ill.: American Nuclear Society.
Holden, Constance. 1990. "Briefings—Incredible Lightness of Gyroscopes." *Science* 247 (12 January): 156–57.
———. 1992. "Random Samples—Why Divorce Runs in Families." *Science* 258 (11 December): 1734.
———. 1994a. "Random Samples—Out of the Cold, onto the Newsstand." *Science* 263 (4 February): 606–7.
———. 1994b. "Random Samples—Psychology in Crisis?" *Science* 265 (22 July): 476.
———. 1995. "The Violent Side of Low Cholesterol." *Science* 270 (13 October): 237.
———. 1996a. "Random Samples—Cold Fusion Gets Drubbing in Italian Court." *Science* 272 (26 April): 487.
———. 1996b. "Random Samples—Happiness and DNA." *Science* 272 (14 June): 1591–93.
Holiday, F. W. 1986. *The Goblin Universe.* St. Paul, Minn.: Llewellyn Publications.
Holland, Dwight. 1998. "Beyond Reason at the Top of the World." *Roanoke Times,* 4 January, Horizon sec., p. 4.
Holmes, Oliver Wendell. 1842. "Homeopathy and Its Kindred Illusions." Lectures to the Boston Society for the Diffusion of Useful Knowledge; reprinted in *Medical Essays,* vol. 9, *Works of Oliver Wendell Holmes* (Boston: Houghton, Mifflin, 1892).
Horgan, John. 1996. *The End of Science: Facing the Limits of Knowledge in the Twilight of the Scientific Age.* Reading, Mass.: Addison-Wesley.
Horrobin, David. 1981. "Nucleus or Nucleolus: Which Runs the Cell?" *New Scientist,* 29 January, pp. 266–69.
Howard, Keith. 1991. "Technology—Magnetic Fields Raise Fuel Efficiency." *New Scientist,* 20 July, p. 20.
Howes, Ruth, and James Watson Jr. 1981. "Science and Pseudoscience: A Course for the Citizen of the Twenty-First Century." *Journal of College Science Teaching* 11, no. 2 (November): 105–7.
Hoyle, Fred. 1984. *The Intelligent Universe.* New York: Holt, Rinehart, and Winston.
Hoyle, Fred, and N. C. Wickramasinghe. 1978. *Lifecloud.* New York: Harper and Row.
Huff, Darrell. 1954. *How to Lie with Statistics.* New York: Norton.
Hufford, David J. 1982. *The Terror That Comes in the Night.* Philadelphia: University of Pennsylvania Press.
Huizenga, John R. 1992. *Cold Fusion: The Scientific Fiasco of the Century.* Rochester, N.Y.: Rochester University Press.

Huyghe, Patrick. 1992. *Columbus Was Last.* New York: Hyperion.
Hyman, Ray. 1980. "Pathological Science: Towards a Proper Diagnosis and Remedy." *Zetetic Scholar* 6 (July): 31–39.
Infinite Energy. 1996. Advertisement for *Exotic Research Report.* Vol. 2 (no. 10): 38.
Jacobs, David M. 1992. *Secret Life.* New York: Simon and Schuster.
Jahn, Robert G. 1980. "Psychic Research: New Dimensions or Old Delusions?" *Zetetic Scholar* 6 (July): 5–16; transcript of an address at "New Horizons in Science," seventeenth annual briefing, Council for the Advancement of Science Writing, 8 November 1979, Palo Alto, Calif.
———. 1982. "The Persistent Paradox of Psychic Phenomena: An Engineering Perspective." *Proceedings of the IEEE* 70, no. 2 (February): 136–70.
Jahn, Robert G., and Brenda J. Dunne. 1987. *Margins of Reality: The Role of Consciousness in the Physical World.* San Diego: Harcourt Brace Jovanovich.
Jain, P. S., Sudarshan Lal, and Henry H. Bauer. 1979. "Polarography and the Kucera Anomaly." *Journal of Electroanalytical Chemistry* 102:425–27.
James Randi Educational Foundation. 1997. Flyer.
Jefferys, William H. 1990. "Bayesian Analysis of Random Event Generator Data." *Journal of Scientific Exploration* 4:153–69.
———. 1992. "Response to Dobyns." *Journal of Scientific Exploration* 6:47–57.
———. 1995. "On *p*-Values and Chance." *Journal of Scientific Exploration* 9:121–22.
Jefferys, William H., and James O. Berger. 1992. "Ockham's Razor and Bayesian Analysis." *American Scientist* 80 (January–February): 64–72.
Johnson, Phillip E. 1991. *Darwin on Trial.* Washington, D.C.: Regnery Gateway.
Jones, Mark M., David O. Johnston, John T. Netterville, James L. Wood, and Melvin D. Joesten. 1987. *Chemistry and Society.* 5th ed. Philadelphia: Saunders.
Josephson, Brian D. 1977. "Molecule Memories." *New Scientist* 1 November, p. 66.
Jueneman, Frederic B. 1989. "Innovative Notebook—Cold Fustian." *Research and Development,* June, p. 49.
Kalmijn, Ad. J. 1982. "Electric and Magnetic Field Detection in Elasmobranch Fishes." *Science* 218 (26 November): 916–18.
Kaufman, Martin. 1971. *Homeopathy in America.* Baltimore: Johns Hopkins University Press.
Kelley, Charles R. 1965. "Orgonomy since the Death of Reich." *The Creative Process* 5 (July), Special Issue.
Kelly, Edward. 1987. "A Too Willing Suspension of Disbelief." *Skeptical Inquirer* 11, no. 3 (Spring): 293–96.
Kendrick, Walter. 1997. "Psychiatrist of the Gods?" *New York Times Book Review,* 21 September, pp. 32–33; review of *Carl Gustav Jung,* by Frank McLynn, and of *The Aryan Christ,* by Richard Noll.
Kerr, Richard A. 1991. "The Lessons of Dr. Browning." *Science* 253 (9 August): 622–23.
Kervran, C. Louis. 1972. *Biological Transmutations.* Trans. Michel Abehsera. Binghamton, N.Y.: Swan House.

Kimball, Roger. 1998. "What Next, a Doctorate of Depravity?" *Wall Street Journal,* 5 May, p. A22.

Kirshvink, Joseph L. 1984. Quoted in "Researcher Links Whale Beachings to Earth's Magnetism," *Roanoke Times and World-News,* 5 December 1984, p. A5.

Knight, David. 1986. *The Age of Science.* Oxford: Basil Blackwell.

Kolin, Alexander. 1968. "Magnetic Fields in Biology." *Physics Today* 39 (November): 39–50.

Kónya, J. 1979. "Notes on the 'Non-Faraday' Electrolysis of Water." *Journal of the Electrochemical Society* 126 (January): 54–56.

Koonin, Steven. 1989. Cited in Ron Dagani, "Fusion Donnybrook: Physicists Assail Utah Claims." *Chemical and Engineering News,* 8 May, pp. 4–5.

Kraatz, Walter C. 1958. "Pseudoscience and Antiscience in an Age of Science." *Ohio Journal of Science* 58, no. 5 (September): 261–69.

Kraft, John C., and Jay F. Custer. 1985. "The Holly Oak Pendant." *Science* 227 (18 January): 244–46.

Krupp, E. C. 1978b. "The Stonehenge Chronicles." In *In Search of Ancient Astronomies,* ed. E. C. Krupp, 81–132, Garden City, N.Y.: Doubleday.

———. 1978c. "Ley or Nay?" In *In Search of Ancient Astronomies,* ed. E. C. Krupp, 253–60, Garden City, N.Y.: Doubleday.

———. 1997. "Archaeoastronomy." In *History of Astronomy: An Encyclopedia,* ed. John Lankford, 21–30. New York: Garland.

———. 1998. Personal communication. 4 August.

———, ed. 1978a. *In Search of Ancient Astronomies.* Garden City, N.Y.: Doubleday.

Kuhn, Thomas S. 1970. *The Structure of Scientific Revolutions.* 2d ed. Chicago: University of Chicago Press.

Kurtz, Paul. 1978. "*The Humanist's* Crusade against Parapsychology: A Discussion." *Journal of the American Society for Psychical Research.* 72 (October): 349–57.

———, ed. 1985. *A Skeptic's Handbook of Parapsychology.* Buffalo, N.Y.: Prometheus.

Kurzweil, Edith. 1998. Editorial. *Partisan Review* 2; abstracted on Internet in *The Monday Review,* 4 May 1998.

Lakatos, Imre. 1976. "History of Science and Its Rational Reconstruction." In *Method and Appraisal in the Physical Sciences,* ed. Colin Howson, 1–40. Cambridge: Cambridge University Press.

Langmuir, Irving. 1968. "Pathological Science." Report no. 68-C-035, General Electric R&D Center, April; transcribed by R. N. Hall.

———. 1985. "Pathological Science: Scientific Studies Based on Non-Existent Phenomena." *Speculations in Science and Technology* 8 (no. 2): 77–94; Langmuir (1968) re-edited with changed Epilogue and added references but without question-and-answer section.

———. 1989. "Pathological Science." *Physics Today,* October, 36–48. Contains shorter revised Epilogue with fewer references than in Langmuir (1985), and no question-and-answer section.

Lankford, John, ed. 1997. *History of Astronomy—An Encyclopedia.* New York: Garland.

Lardner, James. 1988. "Onward and Upward with the Arts—the Authorship Question." *New Yorker,* 11 April, pp. 87–106.

Larson, Dewey B. 1965. *New Light on Space and Time.* Portland, Ore: North Pacific.

———. 1979. *Nothing but Motion.* Portland, Ore.: North Pacific.

Lasswitz, Kurd. 1897. *Auf Zwei Planeten.* Leipzig: B. Elischer Nachfolger; abridged and translated as *Two Planets* (Carbondale: Southern Illinois University Press, 1971).

Lauritsen, John. 1993. *The AIDS War: Propaganda, Profiteering and Genocide from the Medical-Industrial Complex.* New York: Pagan Press, Asklepios.

Leahey, Thomas Hardy, and Grace Evans Leahey. 1984. *Psychology's Occult Doubles: Psychology and the Problem of Pseudoscience.* Chicago: Nelson Hall.

Lefkowitz, Mary. 1996. *Not Out of Africa.* New York: New Republic.

LeGrand, H. E., and Wayne E. Boese. 1975. "*Chariots of the Gods?* and All That: Pseudo-History in the Classroom." *History Teacher* 8, no. 3 (May): 359–70.

Lehn, W. H., and I. Schroeder. 1981. "The Norse Merman as an Optical Phenomenon." *Nature* 289 (29 January): 362–66.

Leo, John. 1997. "Reporters Frame Issues with Feelings, Not Facts." *Conservative Chronicle,* 7 May, p. 19.

Levi, Barbara G. 1973. "Anomalous Water: An End to the Anomaly." *Physics Today,* October, pp. 19–20.

Lewenstein, B. V., and W. Baur. 1991. "A Cold Fusion Chronology." *Journal of Radioanalytical and Nuclear Chemistry, Articles* 152 (no. 1): 273–98.

Lewenstein, Bruce V. 1992. "Cold Fusion Saga: Lesson in Science." *Forum for Applied Research and Public Policy* 7, no. 3 (December): 67–77.

Lewin, Roger. 1981. "A Response to Creationism Evolves." *Science* 214 (6 November): 635–36, 638.

Licht, Sidney. 1959. "History of Electrotherapy." In *Therapeutic Electricity and Ultraviolet Radiation,* ed. Sidney Licht, 1–69. Vol. 4, Physical Medicine Library. New Haven, Conn.: Elizabeth Licht.

Ligon, Woodfin V., Jr. 1989. "Reader Views on Cold Fusion." *Chemical and Engineering News,* 8 May, pp. 2–3.

Lindley, David. 1990. "The Embarrassment of Cold Fusion." *Nature* 344 (29 March): 375–76.

Lindsay Publications. 1990. Catalog no. 505, Winter.

Livingston, James D. 1998. "Magnetic Therapy—Plausible Attraction?" *Skeptical Inquirer* 22, no. 4 (July–August): 25–30, 58.

Love, Dave. 1989. "Cold Fusion and Pseudoscience." *British and Irish Skeptic* 3, no. 4 (July–August): 6–8.

Lowen, Alexander. 1994. *Bioenergetics.* New York: Penguin. First published 1975.

Luben, Richard A. 1995. "Membrane Signal-Transduction Mechanisms and Biological Effects of Low-Energy Electromagnetic Fields." In *Electromagnetic Fields—Biological Interactions and Mechanisms,* ed. Martin Blank, 437–50. Advances in Chemistry, no. 250. Washington, D.C.: American Chemical Society.

Macbeth, Norman. 1971. *Darwin Retried: An Appeal to Reason.* Boston: Gambit.
MacDougall, Curtis D. 1983. *Superstition and the Press.* Buffalo, N.Y.: Prometheus.
Mackie, Euan. 1977a. *Science and Society in Prehistoric Britain.* London: Elek.
———. 1977b. *The Megalith Builders.* Oxford: Phaidon.
Mackintosh, N. J., ed. 1995. *Cyril Burt: Fraud or Framed?* Oxford: Oxford University Press.
Maddox, John. 1989. "What to Say about Cold Fusion." *Nature* 338 (27 April): 701.
Maddox, John, James Randi, and Walter W. Stewart. 1988. "'High-Dilution' Experiments a Delusion." *Nature* 334 (28 July): 287–90.
Mallove, Eugene F. 1991. *Fire from Ice: Searching for the Truth behind the Cold Fusion Furor.* New York: John Wiley.
———. 1996. "Cold Fusion Vindication?" *Chemical and Engineering News,* 17 June, pp. 4, 94.
———. 1997. "Cold Fusion: The 'Miracle' Is No Mistake." *Analog Science Fiction and Fact* 117, nos. 7–8 (July–August): 53–73.
Malmivuo, Jaakko, and Robert Plonsey. 1995. *Bioelectromagnetism: Principles and Applications of Bioelectric and Biomagnetic Fields.* New York: Oxford University Press.
Mangan, Katherine S. 1994. "A&M's 'Alchemy Caper.'" *Chronicle of Higher Education,* 19 January, p. A19.
Mann, Charles. 1990. "Meta-Analysis in the Breech." *Science* 249 (3 August): 476–80.
Mann, Golo. 1938. *Reminiscences and Reflections: A Youth in Germany.* Trans. Krishna Winston. New York: Norton.
Mann, W. Edward, and Edward Hoffman. 1980. *The Man Who Dreamed of Tomorrow.* Los Angeles: J. P. Tarcher.
Marin, Peter. 1982. "What Was Reich's Secret?" *Psychology Today,* September, pp. 56–65.
Markov, Marko S. 1994. "Biological Effects of Extremely Low Frequency Magnetic Fields." In *Biomagnetic Stimulation,* ed. Shoogo Ueno, 91–103. New York: Plenum.
Marks, David, and Richard Kammann. 1980. *The Psychology of the Psychic.* Buffalo, N.Y.: Prometheus.
Marshall, Eliot. 1986. "Yellow Rain Evidence Slowly Whittled Away." *Science* 233 (4 July): 18–19.
Marshall, John C. 1997. "Everyday Tales of Ordinary Madness." *Nature* 389 (4 September): 29.
Martin, David, and Alastair Boyd. 1999. *Nessie—The Surgeon's Photograph Exposed.* East Barnet, Eng.: Martin and Boyd.
Martin, Jim. 1994. "Einstein and the Reichians." *Skeptic* 2 (#4): 14.
Martin, Josef (pseudonym of Henry H. Bauer). 1988. *To Rise above Principle: The Memoirs of an Unreconstructed Dean.* Urbana: University of Illinois Press.
Martin, Michael. 1971. "The Use of Pseudo-Science in Science Education." *Science Education* 55 (no. 1): 53–56.
Mathisen, James A. 1989. "A Further Look at 'Common Sense' in Introductory Sociology." *Teaching Sociology* 17 (July): 307–15.

Matthews, Robert. 1996. "Do Galaxies Fly through the Universe in Formation?" *Science* 271 (9 February): 759.
———. 1997. "The Science of Murphy's Law." *Scientific American,* April, pp. 88–91.
———. 1998. "The Great Health Hoax." *Sunday Telegraph* (London), 13 September.
———. 1999. "Significance Levels for the Assessment of Anomalous Phenomena." *Journal of Scientific Exploration* 13:1–7.
Maultsby, Maxie C., Jr. 1975. *Help Yourself to Happiness through Rational Self-Counseling.* New York: Institute for Rational-Emotive Therapy.
Mauskopf, S. H., and M. R. McVaugh. 1980. *The Elusive Science.* Baltimore: John Hopkins University Press.
McCallum, Dennis, ed. 1996. *The Death of Truth.* Minneapolis: Bethany House.
McClenon, James. 1984. *Deviant Science: The Case of Parapsychology.* Philadelphia: University of Pennsylvania Press.
McDonald, Kim A. 1998. "New Evidence Challenges Traditional Model of How the New World Was Settled." *Chronicle of Higher Education,* 13 March, pp. A22–23.
McHugh, Paul R. 1994. "Psychotherapy Awry." *American Scholar* 63, no. 1 (Winter): 17–30.
Meaden, George Terence. 1991. *The Goddess of the Stones: The Language of the Megaliths.* London: Souvenir.
———. 1997. *Stonehenge: The Secret of the Solstice.* London: Souvenir. Rev. ed. of *The Stonehenge Solution,* 1992.
Medalia, Nahum Z., and Otto N. Larsen. 1958. "Diffusion and Belief in a Collective Delusion: The Seattle Windshield Pitting Epidemic." *American Sociological Review* 23:180–86.
Meecham, Iain A. 1976. "Glint in the Eye for NESSIE." *Amateur Photographer* (London), 21 January.
Meltzer, David J. 1997. "Monte Verde and the Pleistocene Peopling of the Americas." *Science* 276 (2 May): 754–55.
Meredith, Dennis. 1976. "The Loch Ness Press Mess." *Technology Review* 78, no. 5 (March–April): 10–12.
Merton, Robert K. 1957. "Priorities in Scientific Discovery." *American Sociological Review* 22 (no. 6): 635–59.
———. 1961. "Singletons and Multiples in Scientific Discovery." *Proceedings of the American Philosophical Society* 105 (no. 5): 470–86.
Michell, John. 1975. *The View over Atlantis.* London: Garnstone.
Miele, Frank. 1997. "Keeping a Skeptical Eye on Extremists—and on Those Who Keep an Eye on Them." *The Skeptic* 5 (no. 2): 105; review of *American Extremists: Militias, Supremacists, Klansmen, Communists, and Others,* by John George and Laird Wilcox.
Milazzo, G., and Martin Blank. 1983. *Bioelectrochemistry I: Biological Redox Reactions.* Proceedings of the Course on Bioelectrochemistry, Eleventh International School of Biophysics, November–December 1981, Erice, Sicily. New York: Plenum.

Milton, G. W. 1973. "Self-Willed Death or the Bone-Pointing Syndrome." *Lancet,* 23 June, 1435–36.

Morell, Virginia. 1997. "Microbiology's Scarred Revolutionary." *Science* 276 (2 May): 699–702.

Morrill, Ann R. 1990. "Chemical Sensitivity." *Chemical and Engineering News,* 24 September, p. 3.

Moss, Thelma. 1974. *The Probability of the Impossible: Scientific Discoveries and Explorations in the Psychic World.* Los Angeles: J. P. Tarcher.

Mullins, J. M., T. A. Litovitz, and C. J. Montrose. 1995. "The Role of Coherence in Electromagnetic Field-Induced Bioeffects: The Signal-to-Noise Dilemma." In *Electromagnetic Fields—Biological Interactions and Mechanisms,* ed. Martin Blank, 319–38. Advances in Chemistry, no. 250. Washington, D.C.: American Chemical Society.

Needham, Joseph. 1969. *The Grand Titration: Science and Society in East and West.* London: George Allen and Unwin.

Neher, Andrew. 1980. *The Psychology of Transcendence.* Englewood Cliffs, N.J.: Prentice-Hall.

New Times (Los Angeles). 1997. Vol. 2, no. 10 (9–15 October): 8 (advertisement).

NIH GUIDE. 1997. Volume 26, no. 12, 11 April; no. 13, 25 April; no. 23, 18 July; no. 27, 15 August; no. 32, 26 September.

Nordenström, Björn E. W. 1983. *Biologically Closed Electric Circuits: Clinical, Experimental and Theoretical Evidence for an Additional Circulatory System.* Stockholm: Nordic Medical Publications.

North, John. 1996. *Stonehenge: A New Interpretation of Prehistoric Man and the Cosmos.* New York: Free Press.

NOVA. 1991. Television program, "ConFusion in a Jar," first broadcast 30 April.

Nuland, Sherwin B. 1998. "The Uncertain Art: Prooemium." *American Scholar* 67, no. 2 (Spring): 139–42.

Nye, Mary Jo. 1980. "N-rays: An Episode in the History and Psychology of Science." *Historical Studies in the Physical Sciences* 11 (pt. 1): 125–56.

Nyenhuis, John, Joe Bourland, Gabriel Mouchawar, Leslie Geddes, Kirk Foster, Jim Jones, William Schoenlein, George Graber, and Tarek Elabbady. 1994. "Magnetic Stimulation of the Heart and Safety Issues in Magnetic Resonance Imaging." In *Biomagnetic Stimulation,* ed. Shoogo Ueno, 75–89. New York: Plenum.

Oberg, James E. 1986–87. "The Great East Coast UFO of August 1986." *Skeptical Inquirer* 11, no. 2 (Winter): 159–63.

O'Conner, Patricia T. 2000. "Party Girl." *New York Times Book Review,* 23 January; review of *Georgiana: Duchess of Devonshire,* by Amanda Foreman.

Oderwald, Richard G. 1992. "Fusion Feudists." *American Scientist* 80 (March–April): 107.

Officer, Charles, and Jake Page. 1996. *The Great Dinosaur Extinction Controversy.* Reading, Mass.: Addison-Wesley.

ORBL (Orgone Biophysical Research Laboratory) 1998a. Advertisement. *Chronicle of Higher Education,* 20 March, p. A25.

———. 1998b. Flyer announcing Greensprings Educational Programs, Summer, 4 pp.
Ord-Hume, Arthur W. J. G. 1977. *Perpetual Motion—The History of an Obsession.* New York: St. Martin's.
Otten, Alan L. 1985. "People Will Believe Anything, Which Is Why Csicops Exist." *Wall Street Journal,* 24 July.
Palca, Joseph. 1989. "Sun Dagger Misses Its Mark." *Science* 244 (30 June): 1538.
Paton, Alan. 1965. *South African Tragedy: The Life and Times of Jan Hofmeyr.* New York: Scribner's.
Pease, John. 1981. "Sociology and the Sense of the Commoners." *American Sociologist* 16:257–71.
Peat, F. David. 1989. *Cold Fusion.* Chicago: Contemporary Books.
Perry, Thomas. 1997. *Dance for the Dead.* New York: Ballantine.
Peters, Douglas P., and Stephen J. Ceci. 1980. "A Manuscript Masquerade." *The Sciences,* September, pp. 16–19, 35.
Phillips, John B. 1986. "Two Magnetoreception Pathways in a Migratory Salamander." *Science* 233 (15 August): 765–67.
Piccardi, Giorgio. 1962. *The Chemical Basis of Medical Climatology.* Springfield, Ill.: Charles C. Thomas.
Pilla, Arthur A. 1974. "Electrochemical Information Transfer at Living Cell Membranes." *Annals of the New York Academy of Sciences* 238:149–70.
P.N. 1979. "Pseudoscience." In *The Encyclopedia of Science Fiction,* ed. Peter Nicholls, 478–80. London: Granada.
Polidoro, Massimo. 1997. "Secrets of a Russian Psychic." *Skeptical Inquirer* 21, no. 4 (July–August): 44–47.
Pool, Robert. 1989. "How Cold Fusion Happened—Twice!" *Science* 244 (28 April): 420–23.
———. 1990. "Is There an EMF-Cancer Connection?" *Science* 249 (7 September): 1096–98.
———. 1993a. "Evidence for Homosexuality Gene." *Science* 261 (16 July): 291–92.
———. 1993b. "Alchemy Altercation at Texas A&M." *Science* 262 (26 November): 1367.
———. 1994. "Can Sound Drive Fusion in a Bubble?" *Science* 266 (16 December): 1804.
Powell, Mike R. 1998. "Magnetic Water and Fuel Treatment: Myth, Magic, or Mainstream Science?" *Skeptical Inquirer* 22, no. 1 (January–February): 27–31, 63.
Price, Derek de Solla. 1975. *Science since Babylon.* Enlarged ed. New Haven: Yale University Press.
———. 1986. *Little Science, Big Science . . . and Beyond.* New York: Columbia University Press.
Price, R., P. E. Green Jr., T. J. Goblick Jr., R. H. Kingston, L. G. Kraft Jr., G. H. Pettengill, R. Silver, and W. B. Smith. 1959. "Radar Echoes from Venus." *Science* 129 (20 March): 751–53.
Randi, James. 1983a. "The Project Alpha Experiment: Part 1. The First Two Years." *Skeptical Inquirer* 7, no. 4 (Summer): 24–33.

———. 1983b. "The Project Alpha Experiment: Part 2. Beyond the Laboratory." *Skeptical Inquirer* 8, no. 1 (Fall): 36–45.

Rawlins, Dennis. 1981. "sTARBABY." *Fate,* October (reprinted as 32-page booklet).

Reader's Digest Association. 1977. *The World's Last Mysteries.* Pleasantville, N.Y.: Reader's Digest Association.

———. 1981. *Into the Unknown.* Pleasantville, N.Y.: Reader's Digest Association.

———. 1982. *Mysteries of the Unexplained.* Pleasantville, N.Y.: Reader's Digest Association.

Reed, Graham. 1988. *The Psychology of Anomalous Experience.* Buffalo, N.Y.: Prometheus.

Reese, K. M. 1976. "Newscripts—Strong Electric Fields Spur Growth of Chicks." *Chemical and Engineering News,* 26 January, p. 40.

———. 1987. "Newscripts—Bill of the Platypus Has Electroreceptors." *Chemical and Engineering News,* 18 May, p. 92.

———. 1992. "Magnetic Fuel Treatment in Taiwan and Britain." *Chemical and Engineering News,* 21 September, p. 52 (quoting *Asia Magazine* of 7–9 August 1992).

———. 1995. "Newscripts—Cautions about Dietary Data and Recommendations." *Chemical and Engineering News,* 23 October, p. 88.

———. 1996. "Seminar Leaves Magnetic Scale Control up in the Air." *Chemical and Engineering News,* 24 June, p. 96.

Reich, Ilse Ollendorff. 1969. *Wilhelm Reich: A Personal Biography.* New York: St. Martin's.

Reiter, Russel J. 1995. "Melatonin Suppression by Time-Varying and Time-Invariant Electromagnetic Fields." In *Electromagnetic Fields—Biological Interactions and Mechanisms,* ed. Martin Blank, 451–65. Advances in Chemistry, no. 250. Washington, D.C.: American Chemical Society.

Rhodes, Richard. 1997. "Pathological Science." *New Yorker,* 1 December, pp. 54–59.

RJC. 1973. "No Gravity-Wave Confirmation by the New Experiments." *Physics Today,* October, pp. 17–18.

Roberts, David. 1982. *Great Exploration Hoaxes.* San Francisco: Sierra Club.

Rockwell, Theodore, Robert Rockwell, and W. Teed Rockwell. 1978a. "Irrational Rationalists: A Critique of *The Humanist's* Crusade against Parapsychology." *Journal of the American Society for Psychical Research* 72 (January): 23–34

———. 1978b. "Part Two: Context vs. Meaning: A Reply to Dr. Kurtz." *Journal of the American Society for Psychical Research* 72 (October): 357–64.

Root-Bernstein, Robert S. 1993. *Rethinking AIDS: The Tragic Cost of Premature Consensus.* New York: Free Press.

Rosenberg, Alexander. 1995. *Philosophy of Social Science.* Rev. 2d ed. Boulder, Colo.: Westview.

Rostand, Jean. 1960. *Error and Deception in Science.* Trans. A. J. Pomerans. New York: Basic Books.

Roush, Wade. 1997. "Herbert Benson: Mind-Body Maverick Pushes the Envelope." *Science* 276 (18 April): 357–59.

Rousseau, Denis L. 1992. "Case Studies in Pathological Science." *American Scientist* 80 (January–February): 54–63.
Rowbottom, Margaret, and Charles Susskind. 1984. *Electricity and Medicine—History of Their Interaction.* San Francisco: San Francisco Press.
Roy, Rustum. 1989. "Views on Nuclear Fusion." *Chemical and Engineering News,* 15 May, p. 2.
———. 1992. Quoted in Stu Borman, "Solid-State Process Produces Synthetic Diamond," *Chemical and Engineering News,* 26 October, p. 5.
Russell, Bertrand. 1956. "In the Company of Cranks." *Saturday Review,* 11 August, pp. 7–8.
Schiff, Michel. 1995. *The Memory of Water: Homeopathy and the Battle of Ideas in the New Science.* London: Thorsons.
Seife, Charles. 1997. "New Test Sizes Up Randomness." *Science* 276 (25 April): 532.
Selvin, Paul. 1991. "The Raging Bull of Berkeley." *Science* 251 (25 January): 368–70.
Service, Robert F. 1996. "Superconductivity Turns 10." *Science* 271 (29 March): 1804–6.
Shapiro, Arthur K., and Elaine Shapiro. 1997. *The Powerful Placebo—From Ancient Priest to Modern Physician.* Baltimore: Johns Hopkins University Press.
Shapiro, Robert. 1986. *Origins: A Skeptic's Guide to the Creation of Life on Earth.* New York: Summit.
Sharaf, Myron R. 1983. *Fury on Earth: A Biography of Wilhelm Reich.* New York: St. Martin's.
Shattuck, Roger. 1998. "Defending the Middle Class." *New York Times Book Review,* 1 February, p. 11; review of *Pleasure Wars,* by Peter Gay.
Shaw, George Bernard. 1944. *Everybody's Political What's What.* New York: Dodd, Mead.
———. 1946. "Maxims for Revolutionists—Reason," in *Man and Superman.* Harmondsworth, Eng.: Penguin.
Shea, Christopher. 1998. "Tribal Skirmishes in Anthropology." *Chronicle of Higher Education,* 11 September, pp. A17–20.
Sheaffer, Robert. 1998. "Psychic Vibrations—e-mailed Antigens . . ." *Skeptical Inquirer* 22, no. 1 (January–February): 19.
Sherman, J. H., Jr. 1987. "How Can Lang Block Huntington NAS Membership?" (letter). *Research and Development,* September, p. 122.
Shermer, Michael. 1997. *Why People Believe Weird Things: Pseudoscience, Superstition and Other Confusions of Our Time.* New York: Freeman.
Shiels, Tony "Doc." 1990. *Monstrum! A Wizard's Tale.* London: Fortean Tomes.
Shilts, Randy. 1988. *And the Band Played On: Politics, People, and the AIDS Epidemic.* New York: Penguin.
Shuker, Karl P. N. 1995. *In Search of Prehistoric Survivors.* London: Blandford.
Singal, Rajendar K. 1989. "Reader Views on Cold Fusion." *Chemical and Engineering News,* 8 May, p. 3.
SkyMall Magazine. 1998. Advertisement, p. 158.
Small, Christopher. 1973. *Mary Shelley's "Frankenstein"—Tracing the Myth.* Pittsburgh:

University of Pittsburgh Press.. First published as *Ariel Like a Harpy: Mary and "Frankenstein"* (London: Victor Gollancz, 1972).
Smith, Richard D. 1998. Personal communication.
Sobran, Joseph. 1997. *Alias Shakespeare: Solving the Greatest Literary Mystery of All Time.* New York: Free Press.
Society for Scientific Exploration. n.d. See Web page at <http://www.scientificexploration.org>.
Sofaer, Anna, Volker Zinser, and Rolf M. Sinclair. 1979. "A Unique Solar Marking Construct." *Science* 206 (19 October): 283–91.
Spencer, Baldwin, and F. J. Gillen. 1968. *The Native Tribes of Central Australia.* New York: Dover.
Stanley, Steven M. 1987. *Extinction.* New York: Scientific American.
Steele, Shelby. 1990. *The Content of Our Character.* New York: St. Martin's.
Stein, Gordon. 1987. *American Rationalist* 32:44; review of Bauer (1986).
———. 1993. *Encyclopedia of Hoaxes.* Detroit: Gale Research.
Stent, Gunther. 1972. "Prematurity and Uniqueness in Scientific Discovery." *Scientific American,* December, pp. 84–93.
Stephan, Paula E., and Sharon G. Levin. 1992. *Striking the Mother Lode in Science: The Importance of Age, Place, and Time.* New York: Oxford University Press.
Stevenson, Adlai. 1954. Address at Princeton University, 22 March, quoted in *Dictionary of Quotations,* ed. Bergen Evans, no. 22, p. 558. New York: Avenel, 1978.
Stine, O. Colin, Jianfeng Xu, Rebecca Koskela, Francis J. McMahon, Michele Gschwend, Carl Friddle, Chris D. Clark, Melvin G. McInnis, Sylvia G. Simpson, Theresa S. Breschel, Eva Vishio, Kelly Riskin, Harriet Feilotter, Eugene Chen, Susan Shen, Susan Folstein, Deborah A. Meyers, David Botstein, Thomas G. Marr, and J. Raymond DePaulo. 1995. *American Journal of Human Genetics* 57:1384–94.
Storms, Edmund. 1996. "Review of the 'Cold Fusion' Effect." *Journal of Scientific Exploration* 10:185–243.
Straub, Richard E., Charles J. MacLean, F. Anthony O'Neill, John Burke, Bernadette Murphy, Fiona Duke, Rosemarie Shinkwin, Bradley T. Webb, Jie Zhang, Dermot Walsh, and Kenneth S. Kendler. 1995. "A Potential Vulnerability Locus for Schizophrenia on Chromosome 6p24-22: Evidence for Genetic Heterogeneity." *Nature Genetics* 11:287–92.
Sturrock, P. A. 1994. "Applied Scientific Inference." *Journal of Scientific Exploration* 8:491–508.
Sturtevant, William C., and David J. Meltzer. 1985. "The Holly Oak Pendant." *Science* 227 (18 January): 242–44.
Sulloway, Frank J. 1996. *Born to Rebel: Birth Order, Family Dynamics, and Creative Lives.* New York: Pantheon.
Sutton, M. A. 1976. "Spectroscopy and the Chemists: A Neglected Opportunity?" *Ambix* 23 (Pt. 1, March): 16–26.
Szent-Györgyi, Albert. 1971. "Looking Back." *Perspectives in Biology and Medicine* 15, no. 1 (Autumn): 1–5.

Taubes, Gary. 1993. *Bad Science: The Short Life and Weird Times of Cold Fusion.* New York: Random House.

———. 1996. "Rare Sightings Beguile Physicists." *Science* 272 (26 April): 474–76.

Tey, Josephine. 1951. *The Daughter of Time.* London: Peter Davies.

Thomas, David E. 1997. "Hidden Messages and the Bible Code." *Skeptical Inquirer* 21, no. 6 (November–December): 30–36.

Thomas, Lewis. 1979. *The Medusa and the Snail,* New York: Viking.

———. 1981. "Debating the Unknowable." *Atlantic Monthly,* July, pp. 49–52.

Thompson, Damian. 1997. *The End of Time—Faith and Fear in the Shadow of the Millennium.* Hanover, N.H.: University Press of New England; quoted in Robert Eisner, "The Big Round One," *New York Times Book Review,* 9 November, p. 9.

Thomson, J. H., G. N. Taylor, J. E. B. Ponsonby, and R. S. Roger. 1961. "A New Determination of the Solar Parallax by Means of Radar Echoes from Venus." *Nature* 190 (6 May): 519–20.

Tippett, Roger. 1983. Lecture at University of Arizona, September, quoted in *Manchester Evening News,* 30 September.

Tobacyk, Jerome J. 1983. "Reduction in Paranormal Belief among Participants in a College Course." *Skeptical Inquirer* 8, no. 1 (Fall): 57–61.

Trefil, James S. 1978. "A Consumer's Guide to Pseudoscience." *Saturday Review,* 29 April, 16–21.

———. 1983. "Grounds for Dismissal: Case Studies of the Rejection of Scientific Claims." Paper presented at Second Annual Meeting, Society for Scientific Exploration, June, Charlottesville, Virginia.

Triplett, William. 1992. "SETI Takes the Hill." *Air and Space.* October–November, pp. 80–86.

Truzzi, Marcello. 1987. "Reflections on 'Project Alpha': Scientific Experiment or Conjuror's Illusion?" *Zetetic Scholar* 12–13 (August): 73–98.

Turner, S. P., and J. H. Turner. 1990. *The Impossible Science.* Newbury Park, Calif.: Sage.

Tynan, Kathleen. 1987. *The Life of Kenneth Tynan.* New York: William Morrow.

Ueno, Shoogo, ed. 1994. *Biomagnetic Stimulation.* New York: Plenum.

Ullman, Montague, and Stanley Krippner, with Alan Vaughan. 1974. *Dream Telepathy: Experiments in Nocturnal ESP.* Baltimore: Penguin.

Urey, Harold C. 1966. "Biological Materials in Meteorites: A Review." *Science* 151 (14 January): 157–66.

Van Allen, James A. 1986. "Myths and Realities of Space Flight." *Science* 232 (30 May): 1075–76.

Velikovsky, Immanuel. 1946. *Cosmos without Gravitation: Attraction, Repulsion, and Circumduction in the Solar System,* no. 4 in the series *Scripta Academica Hierosolymitana.* New York and Jerusalem.

———. 1950. *Worlds in Collision.* New York: Macmillan.

———. 1982. *Mankind in Amnesia.* New York: Doubleday.

Vidal, John. 1991. "A Circle of Theories at the Bottom of the Garden." *Guardian* (London), 2 August, p. 28.

Wade, Nicholas. 1977. "Schism among Psychic-Watchers." *Science* 197 (30 September): 1344.
Walleczek, Jan. 1995. "Magnetokinetic Effects on Radical Pairs: A Paradigm for Magnetic Field Interaction with Biological Systems at Lower Than Thermal Energy." In *Electromagnetic Fields—Biological Interactions and Mechanisms,* ed. Martin Blank, 395–420. Advances in Chemistry, no. 250. Washington, D.C.: American Chemical Society.
Walsh, John Evangelist. 1996. *Unraveling Piltdown: The Science Fraud of the Century and Its Solution.* New York: Random House.
Watson, Lyall. 1973. *Supernature.* Garden City, N.Y.: Anchor.
Weaver, Warren. 1963. *Lady Luck: The Theory of Probability.* Garden City, N.Y.: Anchor.
Welfare, Simon, and John Fairley. 1980. *Arthur C. Clarke's Mysterious World.* London: Collins.
Werth, Barry. 1994. *The Billion Dollar Molecule: One Company's Quest for the Perfect Drug.* New York: Simon and Schuster.
Wescott, Roger. 1973. "Anomalistics: The Outline of an Emerging Area of Investigation." Paper prepared for Interface Learning Systems.
———. 1980. "Introducing Anomalistics: A New Field of Interdisciplinary Study." *Kronos* 5, no. 3 (April): 36–50.
Westrum, Ron. 1978. "Science and Social Intelligence about Anomalies: The Case of Meteorites." *Social Studies of Science* 8:461–93.
———. 1982. "Social Intelligence about Hidden Events." *Knowledge: Creation, Diffusion, Utilization* 3:381–400.
Wheeler, David. 1997. "Aloe Vera Saves a Doctor's Life." *Whole Life Times—The Journal of Holistic Living,* October, p. 2.
Wheeler, J. A. 1979. "Drive the Pseudos out of the Workshop of Science." *New York Review of Books,* 17 May, pp. 40–41.
Whitaker, Julian. 1997. *Dr. Julian Whitaker's Health and Healing—Tomorrow's Medicine Today,* October 1997 (special supplement).
Whole Person, The. 1997. Calendar of events in Southern California for October.
Wilde, Oscar. 1891. Quoted by Peter Haining, "Introduction," *Ancient Mysteries* (New York: Taplinger 1977).
———. 1901. *Intentions.* London: James R. Osgood McIlvaine.
Wildman, Terry. 1996. "Student Success . . . What *Are* the Limiting Factors?" *The Pedagogical Challenge* (Center for Excellence in Undergraduate Teaching, Virginia Polytechnic Institute and State University) 4 (no. 1): 1, 2, 12.
Wilkinson, Richard. 1988. *The Scientist,* 19 September, p. 17.
Williams, Stephen. 1991. *Fantastic Archaeology: The Wild Side of North American Prehistory.* Philadelphia: University of Pennsylvania Press.
Williamson, Tom. 1987. "A Sense of Direction for Dowsers?" *New Scientist,* 19 March, pp. 40–43.
Wilson, Colin. 1981. *The Quest for Wilhelm Reich.* Garden City, N.Y.: Anchor.
Wilson, James Q. 1997. "Keep Social-Science 'Experts' out of the Courtroom." *Chronicle of Higher Education,* 6 June, p. A52.

Wiseman, Richard, and Robert L. Morris. 1995. *Guidelines for Testing Psychic Claimants.* Buffalo: Prometheus.

Wolfe, Alan. 1998. "The Solemn Responsibilities of Book Reviewing." *Chronicle of Higher Education,* 24 April, pp. B4–5.

Wolpert, Lewis. 1993. *The Unnatural Nature of Science.* Cambridge, Mass.: Harvard University Press.

Zaleski, Carol. 1997. "Singles." *New York Times Book Review,* 5 October, p. 32; review of *Hermits: The Insights of Solitude,* by Peter France.

Ziman, John. 1994. *Prometheus Bound: Science in a Dynamic Steady State.* Cambridge: Cambridge University Press.

Zurer, Pamela. 1988. "Researcher Criminally Charged with Fraud." *Chemical and Engineering News,* 25 April, p. 5.

———. 1996. "'Cold Fusion' Device Hits the Market." *Chemical and Engineering News,* 18 November, p. 9.

Index

HENRY H. BAUER is Austrian by birth (1931), Australian by education (1939–56), and American (since 1966) by choice. He is professor emeritus of chemistry and science studies and dean emeritus of arts and sciences at Virginia Polytechnic Institute. Bauer is the author of *Beyond Velikovsky: The History of a Public Controversy* (1984), *The Enigma of Loch Ness: Making Sense of a Mystery* (1986), *To Rise above Principle: The Memoirs of an Unreconstructed Dean* (under the pseudonym Josef Martin, 1988), and *Scientific Literacy and the Myth of the Scientific Method* (1992).

Typeset in 10.5/13 Minion
with Minion display
Designed by Dennis Roberts
Composed by Jim Proefrock
at the University of Illinois Press
Manufactured by Thomson-Shore, Inc.

University of Illinois Press
1325 South Oak Street
Champaign, IL 61820-6903
www.press.uillinois.edu